T0226810

EMC at Component and PCB Level

EMC at Component and PCB Level

Martin O'Hara

OXFORD AMSTERDAM BOSTON LONDON NEW YORK PARIS
SAN DIEGO SAN FRANCISCO SINGAPORE SYDNEY TOKYO

Newnes
An imprint of Elsevier Science
Linacre House, Jordan Hill, Oxford OX2 8DP
200 Wheeler Road, Burlington, MA 01803

First published 1998
Reprinted 2001, 2003

British Library Cataloguing in Publication Data
A catalogue record for this book is available from the British Library

ISBN 0 7506 3355 7

Library of Congress Cataloguing in Publication Data
A catalogue record for this book is available from the Library of Congress

For information on all Newnes publications
visit our web site at www.newnespress.com

Typeset by ℱ Tek-Art, Croydon, Surrey
Printed and bound by Antony Rowe Ltd, Eastbourne

Contents

Preface

When most electronic circuit design engineers think of electromagnetic compatibility (EMC) they probably think of mains filters, shielded boxes, antennae measurement systems and consider it is mainly the preserve of the radio engineer. This closeted vision of EMC is part of the reason for writing this book, it is not only simplistic, but extremely dangerous to the long-term employment prospects of such thinkers.

At the present time EMC is sending shivers down the spine of all original equipment manufacturers (OEMs), worrying about the prospect of their products being withdrawn from sale in the European Community (EC) market-place due to non-compliance with the EC EMC directive (89/336/EEC). Presently, those people who sell mains filters and shielded boxes are probably making a fortune out of the panic and paranoia surrounding the implementation of this directive. Eventually, it will dawn on some manufacturers that their competitors are producing similar products that are far less expensive than their own and meet EMC directive requirements, without having to install mains line filters or using shielded enclosures, complying due to correct design at the component and printed circuit board (PCB) level.

Whenever a chart is brought out to describe the best and most cost-effective way to produce a product, whether it be for EMC or any other criteria, time and money spent at the earliest design stages always bring about the greatest rewards for the lowest cost. The benefits of implementing correct EMC procedures at the component and PCB stages are not only in the financial gains in production costs, the final equipment will be less expensive hence more competitive, and the time to market will be reduced.

The economics can be examined quite simply by considering any circuit board you have to hand, let's look at a PC as an example. Imagine here are 100 integrated circuits (ICs) on the PC, to decouple every single IC at each device will cost about £1 in total, this may be all that is required to reduce the conducted noise to within the EMC regulations. How much would it cost to fit a mains filter? Most likely in the region of £5. At the PCB level the savings are even greater as there should be no parts cost penalty for following the rules specified in this book. It may even be possible to reduce the parts cost as less decoupling or filtering may be required in the final circuit due to improved PCB layout.

In today's cut-throat, price-sensitive electronics market every penny saved can be the difference between being a market leader and going broke. With cost-effectiveness and time to market prime concerns, there can be no excuse for not following the basic good design practices preached in this book to give your circuit the edge over your competition and improving your chances of meeting the EMC regulatory requirements first time.

Martin O'Hara
1 January 1997

vii

Acknowledgements

This book is a compilation of over 10 years of design experience at the component level coupled with the written experiences and advice of many other workers in the component and electromagnetic compatibility (EMC) fields. I would specifically like to thank the members of Thames Valley EMC Club and its organiser Peter Russell, Helen Crawford of the IEE Library, Steve Jones of Manchester Circuits, Eric Bogatin of Ansoft, and all those companies who kindly provided material for inclusion in this book.

A great deal of this book could not have been completed without the help and assistance of my employer, Newport Components Ltd, who assisted in the use of their measurement equipment and test circuits. I am fortunate in working for a components company who are proactive in the field of EMC and have sponsored several EMC seminars and assisted in the production of the artwork for this book. Special thanks to Dr John Baxter, Lee Frances and Paul Neaves of Newport Components. Finally, thanks to my wife Loraine for hours of understanding and an endless supply of encouragement and tea.

CHAPTER 1

INTRODUCTION

1.1 Electromagnetic Compatibility at Component Level

Component manufacturers are not only exempt from the European electromagnetic compatibility (EMC) directive, but it is actually illegal for them to mark their components with a CE mark claiming compliance with this directive. Similarly, in the USA the Federal Communications Commission (FCC) make no mention of individual component compliance. So why bother looking at the EMC of components?

There are several possible answers depending on your position in the argument. First, for a component supplier it is beneficial to offer an advantage over the competition, it is also possible to charge more for a component with known EMC performance than for one without. Although EMC compliance of components is not a mandatory requirement today, this situation may change and those with measured EMC performance will have a head start. Of course customers may require knowledge of component EMC performance and require data whether it is mandatory or not.

As a consumer of electronic components it is of benefit to know how a part will behave regarding EMC in a system. In the long term it would obviously be preferable to pass any test or performance requirements as far down the electronic food chain as possible (i.e. to the component supplier). In the case where a problem is found, knowing the potential source of the problem can be half-way to fixing it; therefore, the EMC performance of components can be used to trace and eliminate overall circuit or system EMC performance problems.

So whether a supplier or consumer it is beneficial to know the EMC performance and have guidelines on the application of components with regards to EMC. The benefits are both technical and commercial, there is not only a requirement for suppliers to provide the data, but also for consumers to request the relevant data for their application.

The main concern with EMC at component level is the onset of non-ideal behaviour; for example, a capacitor is no longer capacitive above a certain frequency, but resistive or inductive. In general the areas of interest tend to occur at high frequencies, outside the functional operating frequency range of the component. It is operation outside of the functional frequency range due to conducted or radiated electromagnetic interference (EMI) that is the concern of this book. The effects may

1

not be covered by the manufacturer as it is outside the recommended operation; unfortunately, EMI cannot read the data sheet recommendations. Operation in the high frequency area is often outside the applications' information and even outside the experience of many component manufacturers.

1.2 EMC on the Printed Circuit Board

Almost every printed circuit board (PCB) is different and completely application specific. Even within similar products the PCB can be different, for example open two PCs from different manufacturers, with the same processor, clock speed, keyboard interface, etc., the actual PCB layout will be different. This diversity means that every PCB has a unique level of EMC performance, so what can possibly be done to ensure that this is within certain limits?

It should not surprise circuit designers that the layout of the PCB can have a significant effect on the EMC performance of a system, usually more so than the actual choice of components. Consequently, PCB layout is one of the most critical areas of consideration for design to meet EMC regulations.

The fact that there are so many different PCB designs in existence is a testimony to the low cost of producing a PCB, but relaying a complete PCB because of poor layout design causes significant increases in costs not present in the actual material price of the board. Relaying a PCB will create a delay in time to market, hence lost sales revenue. New PCB layouts or changes usually entail new solder masks, reprogramming component placement machines, rewriting the production instructions, etc., hence cost may not be present in the final product part cost, but in the development and production overhead.

Although a significant factor in overall EMC performance, the recommendations for minimising the effect of PCB layout on EMC are general good PCB design practices. The cost of implementing these recommendations is solely in the time taken to ensure that these good design practices are implemented, vigilance and experience are the two main requirements, not necessarily new design software or extensive retraining.

1.3 Parameters Relating to EMC Performance

It is well documented in several other texts on EMC that the parameter which need to be examined are: frequency, amplitude, time, impedance and dimensions. This is sometimes abbreviated to FATID, usually pronounced 'fatted', as in bringing the fatted calf to the slaughter (apologies to any vegetarians reading this).

In components it tends to be frequency, amplitude and impedance that dominate the interest. Usually, the frequency outside the normal range of operation and how much these signals are being attenuated (impeded). Examination of the amplitude of signal required to operate a device is also useful, generally devices with a higher operating threshold have higher immunity than those with low amplitude operating points.

With a PCB it is usually the frequency and physical dimensions that are the dominant parameters of interest. The advantage of a PCB is that as a designer we can exhibit full control over the physical dimensions, unlike with a component where the degrees of freedom are somewhat restricted. The physical dimensions of PCB tracks and interconnect effect the frequencies which the circuit will be susceptible to and which it can radiate best.

The time parameters can be converted to a frequency if the signal is continuous, or if it is the timing of edges that is causing a problem. The other time-dependent effect is determining if the problem is caused by a specific timed action within the circuit or system (e.g. switch action such as energising a relay). It is much more common in EMC to deal with frequency than time and examine signals in the frequency domain.

1.4 What's In It For Me?

It cannot be guaranteed that in following all the design and component suggestions made in this book that every design will pass through EMC testing first time; however, by following the ideas postulated here the chances should be improved. It could be argued that not all the ideas are feasible together and I would not want to make anyone think that they had to follow all these suggestions to achieve EMC compliance. By careful use of some of the ideas, as and when appropriate, and by experience in applying them to your circuit designs, I am confident that the reader of this book will realise better EMC performance from their circuits at virtually no additional cost.

What all readers of this book should be able to achieve is that their circuits are optimised for EMC performance by following best design practices. They should have a better understanding of potential sources of EMC problems in existing circuits and have some idea of how to fix them. The reader should be armed with a design tool kit that allows them to produce the best EMC performance in the most cost-effective manner with the minimum requirements for post-layout add-ons such as mains filtering and shielding.

1.5 Summary

The most cost-effective way of complying with any requirements in a circuit, system or end product is to consider the requirements at the earliest stages of design (Figure 1.1). The aim of this book is to take the focus for EMC all the way down the electronics food chain to the component and PCB level. Designers should be aware at the outset of their design on how the choice of component type and placement of

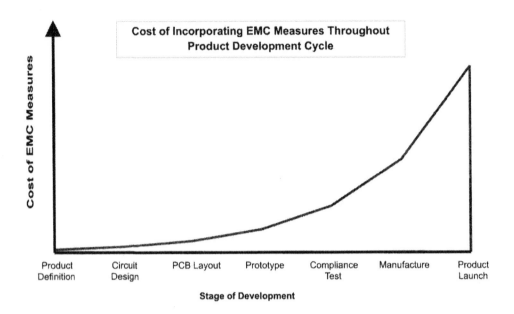

Figure 1.1

Cost of EMC measures

components will affect the EMC of their final circuit, as well as what additional protection may be required. This is of course supplementary to the circuits initial requirement of functional performance.

All component suppliers are exempt from the EU EMC directive and FCC regulations and, therefore, have no legal obligation to demonstrate compliance or issue EMC information. The more enlightened component supplier will already have some information and should be able to help advise of application pitfalls or give guidelines for EMC considerations. Do not be afraid to ask, the more this requirement is asked of suppliers the more likely they are to supply the information.

One problem many suppliers of components and their customers alike have is knowing what information would be useful. The application areas for a resistor, for example, are so diverse as to defy a general statement on 'best' method of application. If you know what information you would like to see get in touch with your component supplier, they may not be able to provide it immediately but by informing the component supplier of the need for certain information should see this eventually being included in the data sheet. Similarly, if one supplier can give you the data whereas another cannot, or will not, this could be a suitable method for reducing vendor selection or changing vendor ratings (e.g. if the impedance analysis of a network transformer is not given by one supplier, but is shown to be suitable by another, why risk a possible EMC problem).

PCB suppliers are a little bit more cognisant with regards to EMC. It has been known for a long time that the layout of a circuit is one of the major influences in the end circuits' noise performance, hence its EMC performance. Even with the increased awareness there are still few suppliers who can offer tightly controlled impedance characteristics. Again there are no legislative obligations on the PCB supplier to provide a quiet PCB and it is ultimately the responsibility of the PCB designer or layout engineer.

It is unlikely that EMC will be the primary concern when first choosing components for a circuit design or when producing a PCB layout. If the advice given here is kept in mind, however, the possibility of poor component choice or PCB layout causing EMC problems should be minimised. After all, EMC begins and ends at the circuit level.

CHAPTER 2

PASSIVE COMPONENTS

The selection of passive component elements in a circuit is often overlooked as these components are usually chosen to bias or complement their more exciting active component counterparts. The passive component does have a significant effect on the overall electromagnetic compatibility (EMC) of a circuit, as these components cannot only be the cause of circuit problems in themselves, but can also upset the stability of the active circuit they are connected to.

Passive component electromagnetic interference (EMI) problems could result in the expensive mistake of replacing a perfectly good active device, such as an op-amp, because a bias resistor is acting inductively at a certain frequency for example. The change could not only be more expensive, the addition of extra filtering to reduce some noise source, but may simply be masking the real problem and not truly solving it.

Care therefore needs exercising in the choice of passive component for certain circuit applications. There will be instances where absolute value is not as important as construction or component material for the EMC performance of the circuit. Few manufacturers like to admit that their component is not suitable for any particular application, it is therefore left up to the designer or production engineer to decide if component choice, say using a carbon instead of metal film resistor, is going to change the EMC performance of the circuit. Without the right background knowledge this type of decision cannot be made correctly.

2.1 Passive Component Packaging

There are essentially only two types of package for all electronic components, these are leaded or leadless. The two package types use different technology for final assembly of the circuit.

Leaded components are mounted on the opposite side to the tracking and the leads pass through the printed circuit board (PCB) to make electrical contact with the circuit. Consequently, leaded components are sometimes referred to as through hole components (due to the common practice of plating through holes on a PCB this type of component is also referred to by the abbreviation PTH component). Leaded technology is the elder method of component attachment, virtually all component types are available in a leaded package. Types of leaded package are relatively numerous, however passive components tend to stick to two form factors; axial or radial. Axial leaded components feature a cylindrical component structure with leads concentric with the component body at either end. Radial leaded components have leads which extend from the base of the component structure.

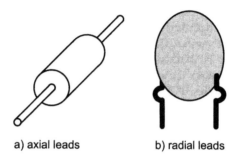

a) axial leads   b) radial leads

Figure 2.1

Through hole components

Leadless components are mounted on the same side as the tracking connecting the component to the circuit, hence these are often referred to, probably more correctly, as surface mount (SM) components. Certainly the term leadless is a slight misnomer as there are of course some form of lead or termination between the component structure and the circuit contacts. Although SM has been around for some time, it is not as universally used as through hole technology on low volume circuits. Primarily due to the cost of the component handling equipment, SM is more commonly found on high volume products and more recent circuit designs requiring a high packing density. Not every type of passive components is available in a SM form, hence there is not always as much choice as with leaded types. Leadless passive components also have a limited variety of package styles, the most dominant is the rectangular body with rectangular end terminations. There are circular bodied types (MELF), but these tend not to be as popular due to handling and placement issues.

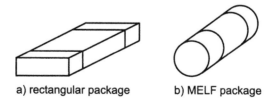

a) rectangular package   b) MELF package

Figure 2.2

Surface mount packages

The package parasitic of leaded components are dominated by the lead length. At high frequencies the lead forms a low value inductor, a typical value for lead inductance is 1 nH/mm per lead (i.e. a component with 10 mm leads will have a parasitic 20 nH inductance in series with it, 10 nH per lead). The end terminations can also produce a small capacitive effect, in the region of 4 pF (based on axial shape with metal end caps on a 10 mm body), but it is usually the lead inductance that is the dominant parasitic component. Consequently, the lead length should be reduced as much as possible. Drilled through holes in the PCB should be spaced just longer than the body length for axial components and the component should be mounted close to the PCB surface. Radial leaded components naturally allow for shorter leads as the body can be maintained flush with the PCB surface and no bend is required for the body diameter (as with axial parts). A radial leaded part can have a lead length almost equal to the PCB thickness. Some lead forming may be required for the manufacturing process (i.e. to stop the parts falling off the PCB during wave soldering), this forming should be made with the shortest possible lead lengths to hold the component body flush to the PCB surface.

Figure 2.3

Loop area of through hole components

SM technology has the shortest lead lengths possible by design, hence there is little the user can do to reduce further the parasitic effects of this package type. There is still a parasitic inductive effect at high frequencies and the main benefit of SM over leaded components is that this is much better controlled and stable, the variations in lead length that can be observed in similar type of leaded parts are not manifest in an SM package. Typically 0.5 nH of parasitic inductance is present in most SM packs with a small end termination capacitance of about 0.3 pF. As with leaded parts it is the parasitic inductance that dominates at high frequency. The different SM sizes (usually quoted by reference numbers, e.g. 1206, 0805) tend to produce reasonably similar parasitic values of inductance and capacitance. Certainly the variations between types are small because as body sizes increase, end terminations also get larger and the net effect is a similar package parasitic.

The different mounting options therefore produce different additional parasitic component elements to be considered. The preferred choices from an EMC viewpoint would be first SM, then radial leaded and lastly axial leaded. At the end of the day it is most likely that component package choice will be down to assembly technology and part type availability, but being aware of some of the EMC issues at the package level can help minimise potential EMC problems.

2.2 Resistors

The simplest element in almost all circuits, surely the good old resistor cannot affect the EMC performance of a circuit? Unfortunately, the wrong choice of resistor type can actually affect several aspects of circuit performance, not only EMC but higher frequency functional performance. The choices are mainly connected with the physical construction of the resistor and their parasitic effect.

2.2.1 Resistor Construction

Leaded resistors are almost exclusively formed in an axial package style. There is generally a choice of three common materials: carbon, metal film or wire wound. Other materials do exist but these are the most common forms, and much of the discussion can be applied to other materials knowing the material properties and the resistor construction.

The construction of film type (metal and carbon film) and wire wound types of leaded resistor are identical, essentially a helical track of the resistive element is placed on to a thermally conductive body (ceramic being the preferred body material). The resistance is governed by the thickness and length of the track. The shape can lead to the potential for a large inductive element, which can be virtually eliminated with tracked film resistors, as the tracks can be made to zigzag rather than perform complete circles. With wire wound resistors this can be performed to some extent by reversing the direction of the helix midway through the winding, but this is not easy to produce and not a common technique.

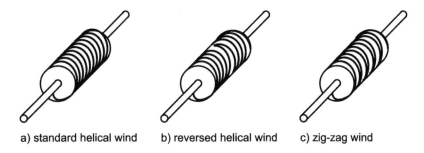

a) standard helical wind   b) reversed helical wind   c) zig-zag wind

Figure 2.4
Film and wire wound resistor construction

The other type of carbon resistor is produced from a solid body of a carbon compound, therefore no tracking is present. This resistor's resistance is affected by the mix of the carbon compound forming the body. Carbon-bodied resistors tend to be lower power rated than metal film or wire wound types as the heat generated by the resistive element in a film or wire type is easily conducted away from the element. With solid-bodied resistors the heat can affect the resistance if allowed to rise to a high enough value.

There is an obvious hierarchy or preference in leaded resistors with regards to EMC performance, with preference for the carbon body types. Wire wound resistors are highly inductive and should be avoided in any frequency sensitive circuit. The shape of a spiral wound wire can be considered as a solenoid, hence the component will not only behave inductively in circuit, but is quite a susceptible design and could introduce signals into the circuit. Metal film tend to have lower inductance but this still dominates as a parasitic element at relatively low frequencies (MHz region). Carbon film is not a particularly good conductor, hence its use as a resistive element, this also reduces its ability to form an inductor. Carbon bodied-leaded resistors are dominated by the parasitic end terminations and leads rather than the resistor construction.

SM leadless resistors come in two basic construction types: thick film and thin film. Thick film types have a resistive layer on a carrier (again usually ceramic) and the film is 'trimmed' along one edge to produce the desired resistance value. Thin film resistors have a 'snake' of resistive film on a carrier forming the resistor. Neither type has a significant inductive parasitic and the end terminations tend to be the dominating parasitic component at high frequencies. Typically in small sizes (0603, 0805) this end termination gives 0.3 pF of capacitance, in larger body sizes (2010, 2512) this can become as low as 0.05 pF. There is some termination inductance, in the region of 0.5 nH, but this is insignificant except at extremely high frequencies. There is no hierarchical preference for either construction of SM resistor, both are preferable to leaded types for their EMC performance.

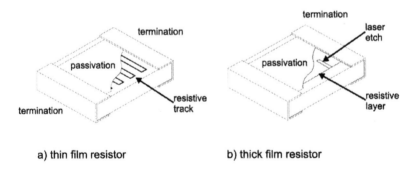

a) thin film resistor   b) thick film resistor

Figure 2.5

Surface mount resistors

2.2.2 Resistance Value

The absolute resistance value can have an effect on the EMC performance. In wire wound types for instance the parasitic inductance will dominate at a lower frequency the more wire that is on the part (i.e. at higher resistance values). Precision wire wound resistors can be constructed by using an equal number of turns and changing the wire gauge to produce the required resistance value, hence equal parasitic inductance and capacitance is present regardless of resistance value.

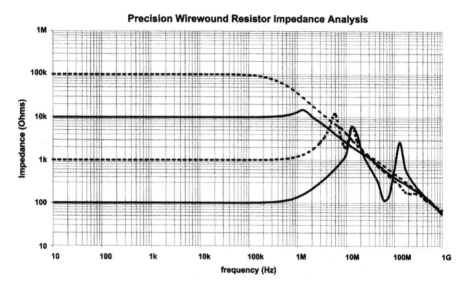

Figure 2.6

Wire wound resistor impedance

With leaded carbon and metal film types the lower the resistance the more prominent the lead inductance, hence at high resistance values lead inductance can be swamped and the end cap terminations become the dominating parasitic element, this only affects the component performance at very high frequencies (>100 MHz).

In SM packages the end termination parasitics dominate regardless of construction. The parasitic capacitance is in the sub-pF range so only becomes noticeable with higher resistance values and at very high frequencies.

Resistors also have a noise voltage associated with their construction. In wire wound types this is extremely small, metal film types have a noise voltage of about 0.1 µV/V (i.e. 0.1 µV of noise is generated for every 1 V of applied bias). Carbon types have a noise voltage dependent on the resistance, given by the equation:

$$V_n = 2 + \log\left[\frac{R}{1000}\right] (\mu V/V) \qquad 2.1$$

The value of noise voltage is small even for large resistor values, for example a 100 k resistor only generates 4 µV/V. In the majority of EMC critical applications it will be the effect of the reactive parasitic elements which will limit performance not the noise voltage. Noise voltages of this level are of more concern for circuits which require a low noise threshold for their intrinsic function (e.g. high precision converters, precision voltage references, audio amplifiers). The frequency of resistor noise voltage also tends to fall as a $1/f$ law, hence this noise voltage has little effect on the EMC performance of a resistor.

2.2.3 EMC Critical Resistor Applications

The most important consideration for the majority of circuit designers is what is the best type for my application? Although the application of resistors is so varied it is next to impossible to answer all applications within any single book, it is possible to generalise on a few circuits by their normal operating frequency.

If possible, SM types are always preferable, these offer the lowest parasitic elements and are relatively stable over temperature and life due to their film on ceramic construction. With through hole parts the carbon body or carbon film types are the best option, with metal film very good where a higher power density or high accuracy is required. Wire wound resistors should only be used where the high power handling of this type is essential. Some of the very high power wire wound types are available in a metal heat sink body cladding, this can be connected to a ground or even one end of the resistor to reduce the inductive effect of the construction and provide some shielding from any internally generated fields.

In gain setting on amplifier circuits the effect of increased impedance due to inductive reactance can cause large differences in the circuit gain at higher frequencies. In high frequency amplifiers the resistor gain setting network needs to be located very close to the amplifier, especially at the input terminal (whether a discrete or integrated amplifier). In audio and lower frequency circuits placement tends not to be quite as critical, but minimising distances between connected circuit components is generally a good practice.

Biasing resistors in general do not have to be of a high tolerance, hence low cost carbon-bodied types are perfectly adequate. If biasing the outputs of a fast switching

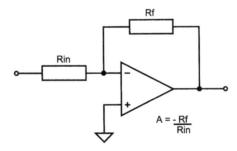

Figure 2.7

Locate gain setting resistors close to amplifier

transistor or integrated circuit (IC) then placement needs to be close to the active component terminals and power (pull-up) or ground (pull-down) connection at the local supply point of the active device. Parasitic inductance on fast-edged switches can create some ringing, and hence requires minimising.

On DC bias points resistor placement and the parasitic lead effects tend not to be as critical. The exception for DC bias circuits are regulator and reference devices which are minimally decoupled to improve the transient response times, therefore will require close placement of the resistors to the active device.

In RC filter networks the wire wound resistor may give a much higher roll-off curve than other types due to its highly inductive nature. Some knowledge of the inductive effect will need to be known if a controlled roll-off is required, otherwise roll-off may occur earlier than expected due to the parasitic inductance. This is one situation in which the inductive parasitic of a wire wound resistor can be used to benefit the EMC performance of the circuit. Care needs to be taken for the possibility of local oscillations if a resonance mode is produced between the parasitic inductance and any capacitor in the circuit, effectively forming a tank circuit, or even within the resistor construction itself (see 100Ω wire wound resistor impedance in Figure 2.6).

Overall, with the exception perhaps of wire wound types, resistors are relatively stable elements. Care should be taken if designing very high frequency circuits that gain roll off is not occurring due to parasitic inductive or capacitive impedance effects rather than the resistance itself.

It is not possible here to cover every resistor type or application, next time your supplier calls ask for further information on the parasitic elements present in the types in use in your circuit. The suppliers are the best source of information and may even be able to change their component construction to suit your application if you require enough resistors of a specified type.

2.3 Capacitors

The capacitor is potentially the easiest and cheapest way to solve many EMC problems. However, the correct choice of capacitor needs careful consideration as not all capacitors behave the same over a wide frequency range. Capacitor selection can be a headache due to the number of types available, their individual behaviour and the possibility of parallel resonance when different values or types are used in the same circuit. Choice should be made on the basis of application and with a good idea of the frequency range over which the capacitor is intended to work.

2.3.1 Construction

The construction of a capacitor is to a certain extent similar for all types, essentially there are two plates of metal with a sandwich of dielectric between them. It is the dielectric material used for charge storage which determines the main capacitor characteristics. There is quite a diversity of constructional methods and so for simplicity only a few of the most common types will be described here. One comforting factor throughout the whole range of dielectric and constructional types is that the same parallel plate capacitor equation is used with all types to calculate the designed capacitor value:

$$C = \varepsilon_0 \varepsilon_r \frac{A}{d}$$

2.2

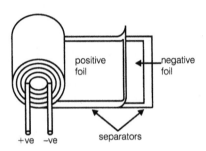

Figure 2.8

Aluminium electrolytic capacitor construction

Where ε_o is the dielectric constant of free space (8.854 pF/m), ε_r is the relative dielectric constant of the dielectric material, A is the overlapping plate area and d is the distance between plates.

Leaded capacitors are available in both axial and radial forms, with radial being the preferred form for most dielectric types. The advantage of radial leads are the ease of producing very short lead lengths on to the PCB, leads can be cropped almost flush with the PCB. Axial-leaded parts should be mounted with the body flush to the

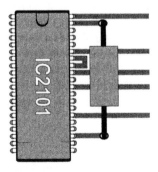

Figure 2.9

Axial decoupling capacitor bridging PCB tracks

PCB and with minimum excess lead lengths. There are situations where an axial leaded part is advantageous, if for example there are tracks which need bridging to decouple an IC, the capacitor body can be used to form the bridge. At high frequencies the inductance of the leads reduces the effectiveness of the capacitor and in the majority of cases radial-leaded components would be the preferred choice of leaded part.

Aluminium electrolytic capacitors are usually constructed by winding metal foils spirally between a thin layer of the dielectric. This constructional method gives high capacitance per unit volume but increases the internal inductance of the part. Tantalum types are manufactured from a block of the dielectric with direct plate and pin connections, and hence have a lower internal inductance than aluminium electrolytic types.

Ceramic capacitors have a construction consisting of multiple paralleled metal plates within a ceramic dielectric (SM types are often referred to as multilayer ceramic chip capacitors, MLCC). The dominant parasitic is the inductance of the plate structure and this usually dominates the impedance for most types in the lower MHz region. The type of ceramic dielectric appears to have little effect on the parasitic inductance (standard ceramic materials have very close natural resonant frequencies: COG; 40 MHz, X7R; 44 MHz, Y5V; 46 MHz). This makes the estimation of useful frequency range relatively easy as a parasitic package inductance of 0.4 nH can be used and the self-resonant frequency (the frequency at which the capacitor has minimum impedance) can be estimated from the nominal capacitance value:

$$f = \frac{1}{2\pi \sqrt{L_p C}} \qquad\qquad 2.3$$

The above equation can be rearranged to calculate the maximum value of capacitance above which parasitic lead inductance limits frequency response. For ceramic capacitors this value is approximately 40 nF, that is below 40 nF ceramic

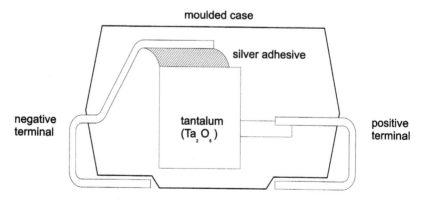

Figure 2.10

Surface mount tantalum capacitor construction

capacitors of Y5V, X7R and COG dielectric are limited by the natural frequency response of the material not by package parasitics.

As with resistors, package type is important and for the same reason, the internal structure and end terminations behave inductively at high frequency. Preference would again be for SM types; however, the choice and range of capacitors available in SM is limited compared with through hole types. More types are becoming available in SM packages and as well as the popular ceramic types, aluminium and tantalum dielectric types are now available in a SM package from many suppliers.

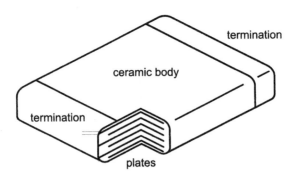

Figure 2.11

Ceramic capacitor

2.3.2 Capacitor Dielectric

The dielectric material used in the construction of a capacitor to some extent defines its application area, each dielectric has a limited frequency range over which it is usable. Aluminium and tantalum electrolytic types dominate at the low frequency end, mainly in reservoir and low frequency filtering applications. In the mid frequency range (kHz to MHz) the ceramic capacitor (multilayer structured types) is the dominant dielectric for decoupling and higher frequency filters. Special low loss ceramic and mica capacitors are available, at a cost, for very high frequency applications and microwave circuits.

The use of ceramic capacitors needs careful consideration, as some of the lower cost ceramic dielectrics (Y5V or Z5U and X7R) have a voltage bias dependence and temperature variation. The effect is a significant fall of capacitance with DC bias voltage, hence the effective capacity available in a circuit may be much less than is required or has been designed for. This may be a problem in decoupling applications, particularly at high bias voltages (12 V and higher), so care should be taken in selection even of ceramic types. Even at 5 V the lowest cost ceramic dielectric (Y5V and Z5U) exhibits a loss of 20% of its nominal capacitance value, this is ignoring any additional tolerance variation and thermal effects.

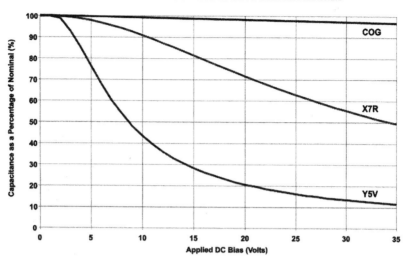

Figure 2.12

Ceramic dielectric bias dependence

There is of course a cost and sometimes a size penalty with better ceramic dielectric. The COG and NPO types are very stable but do not have the storage density of the lower cost materials, and, hence, often a larger body size may be required for the same value of capacitor. For example, using Y5V dielectric may enable a 1 μF capacitor to be produced in an 0805 size body, in X7R the same capacitance value may only be available in a 1206 size and in 1210 for COG. At the values used for decoupling in digital circuit applications (in the nanofarad range) this should not be a great problem as all values can be obtained in small (0805 or less) body sizes. Also the absolute value of capacitance is not quite as critical for high frequency decoupling applications, even a 50% loss of nominal capacitance should still maintain most of the decoupling effectiveness.

2.3.3 Capacitor Impedance Plot

One of the best ways to assess if a capacitor is suitable for an application, with respect to frequency, is its impedance plot. This is a graph of the impedance between the terminals as a function of the applied frequency. The ideal capacitor equation would suggest that the impedance (Z_c) would continue falling indefinitely and is dependent solely on the capacitance (C) for any applied frequency (f):

$$Z_c = \frac{-1}{j2\pi f C}$$
2.4

The actual plot shows the effect of parasitic inductance (L_p) in that the impedance eventually starts rising again. Hence the capacitor exhibits a self-resonant frequency.

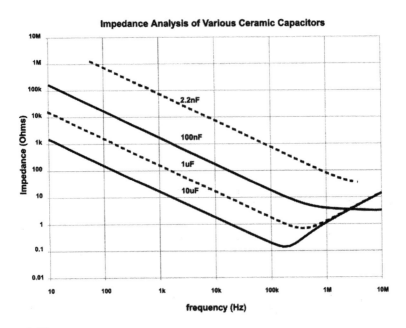

Figure 2.13

Impedance plot of various ceramic capacitors

This is the frequency at which the parasitic inductance and design capacitance have an equal reactance. Self-resonance frequency (f_o) can be calculated, if not quoted in the data sheet, from:

$$f_o = \frac{1}{2\pi \sqrt{L_p C}}$$ 2.5

On some capacitors the lowest value of the impedance plot is not limited by this self-resonance effect, but by the equivalent series resistance (ESR). The ESR is a figure for the effective short circuit ability of the dielectric. The ESR value for ceramic dielectrics is typically under 1Ω, often close to 0.1Ω or lower. Tantalum materials exhibit ESR values in the 1–5Ω range, with aluminium dielectrics having higher ESR values up to 10Ω. The equivalent series resistance value limits the lowest value at which the impedance can fall to, and often disguises the actual self-resonance frequency. For best EMC performance it is important to have a low ESR value as this provides a higher attenuation to signals, especially frequencies close to the self-resonant frequency of the capacitor in use.

The equation for the impedance (Z) can be rewritten to include the parasitic inductive impedance (Z_L) and ESR value if a more detailed analysis is required:

$$Z = Z_L + Z_c + ESR = j2\pi f L_p - \frac{1}{j2\pi f C} + ESR$$ 2.6

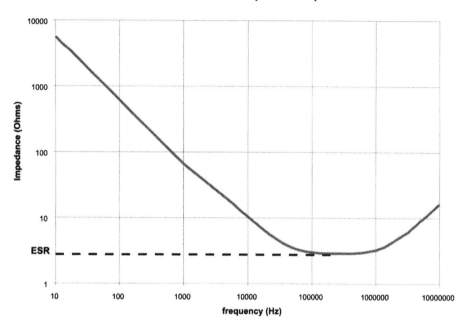

Figure 2.14

ESR limited capacitor

It is generally the self-resonant frequency that limits the useful frequency range of a capacitor, hence the need to limit parasitic inductance. It can also be observed that a higher operating frequency range is obtained from lower value capacitors in the same dielectric and package type until limited by the dielectric material frequency limit.

2.3.4 Bypass Capacitor

One primary use of capacitors in many circuit designs is to act as a high frequency bypass source for switching demands. The bypass capacitors also tend to be used as supply voltage hold-up capacitors and act as a ripple filter to reduce the transient circuit demand on the power supply unit (PSU) directly. The capacitor reduces transient current demands being routed around other circuits attached to the PSU by providing the short demands from the local reservoir bypass capacitor. A slower current demand on the PSU is made by the capacitor for recharging. The current demand is considered as 'bypassing' the PSU and being supplied by the capacitor.

Bypass capacitors tend to be in the range of 10 to 470 µF per PCB or circuit function. The value required can be calculated if the transient current demand (ΔI), the allowable voltage droop (ΔV) and the PSU to PCB lead inductance (L_{PSU}) are known.

The maximum allowable supply impedance (Z_{PSU}) is given by:

$$Z_{PSU} = \frac{\Delta V}{\Delta I}$$ 2.7

The maximum switching frequency the PSU can tolerate (f_{PSU}) without a bypass capacitor is estimated by the ratio of lead impedance to supply impedance:

$$f_{PSU} = \frac{Z_{PSU}}{2\pi L_{PSU}}$$ 2.8

This value usually calculates to be in the 10 kHz to 1 MHz region; therefore, unless the target circuit operates at low speeds (e.g. audio circuits) or on DC only, a bypass capacitor will be required. The bypass capacitor value can then be calculated from the switching frequency available from the supply to match the impedance of the target circuit demand:

$$C_{bypass} = \frac{1}{2\pi f_{PSU} Z_{PSU}}$$ 2.9

For example, consider a PCB containing 50 HC CMOS digital ICs (10 ns rise/fall time), each with four gate drivers into loads of 5 pF per gate operating with a 3.3 V supply rail. The current demand for a simultaneous switch into these capacitive loads is given by:

$$\Delta I = nC \frac{V_{cc}}{\tau_r} = (200) \times (5 \text{ pF}) \times (\frac{3.3}{10 \text{ ns}}) = 0.33 \text{ A}$$ 2.10

The lowest voltage droop allowable is 0.1 V, this is the design safety margin given in the specification for the circuit. The supply impedance to this board is therefore 0.303Ω. The PCB has 10 cm leads to the PSU (approximately 200 nH lead inductance), hence the supply can provide adequate bypass up to a frequency of 241 kHz. If the digital clock exceeds this frequency (most digital circuits use a higher clock frequency than this) then a bypass capacitor of more than 2.2 µF is required.

Note that these calculations give the lowest recommended bypass capacitor value; for the above example a 4.7 µF or 10 µF would provide a better bypass. Often the actual current demand and voltage droop are difficult to gauge and the calculations cannot be made. A general rule of thumb is to use between 10 and 470 µF per PCB, using higher value bypass capacitors for longer leads between PCB and PSU, larger PCBs with higher numbers of ICs and faster rise times circuits. The high value of the bypass capacitor usually excludes ceramic as the dielectric in most bypass applications. Consequently, aluminium and tantalum dielectric capacitors are often used in bypass applications.

2.3.5 Decoupling Capacitor

The values calculated for bypass capacitance can be observed on an impedance plot to not be effective at the frequencies that are typically used for switching most digital ICs (most 10 µF capacitors for instance have a self-resonant frequency about

100 kHz). This dichotomy occurs due to the fact that the bypass is primarily aimed at preventing the local supply from drooping by more than the allowable noise margin and not directly decoupling the switching frequency noise itself. The switching frequency noise needs decoupling to ground by a lower value capacitor, specifically employed to ground out this higher frequency noise and located as close as possible to the source of the noise (i.e. close to each IC or discrete circuit).

The bypass capacitor can also have a decoupling capacitor sited with it at the PCB supply inlet. This helps filter high frequency noise and a value of approximately 1/100–1/1000 of the bypass capacitor is commonly used (e.g. for a 10 μF bypass capacitor a 100 nF or 10 nF decoupling capacitor is placed in parallel at the inlet connection).

Even with a decoupling capacitor in parallel with the bypass there is often insufficient noise rejection from switching at the individual ICs. There is still a relatively large track impedance between the supply inlet decoupling capacitor and each IC or discrete circuit function. There needs to be a decoupling capacitor at each IC and close to any discrete circuit performing a high speed function (e.g. transistor mixer circuits). Consequently, for improved EMC performance there are going to be a number of discrete low value capacitors distributed around the circuit and located close to each IC or discrete switching function.

The traditional method is to use the standard capacitance-current and voltage-time derived equation as used to calculate the current demand in the bypass capacitor example above. The equation for current to/from the capacitor is:

$$i = C\frac{\mathrm{d}V}{\mathrm{d}t}$$

2.11

Rearranging for the capacitance (C) for a given current demand (i), assuming the rise time (τ_r) and allowable noise voltage (ΔV) are known;

$$C = i\frac{\tau_r}{\Delta V}$$

2.12

Take an example of an oscillator circuit with a rise time of 2 ns (regardless of actual oscillator frequency), driving a 50 mA line and as this is in the same circuit as our previous bypass capacitor example, a maximum voltage noise of 100 mV. The calculated decoupling capacitor is 1 nF, again a slightly higher value adds in some margin for error, say 2.2 nF or 4.7 nF. Too large a value may not decouple the harmonics adequately and again reference to the capacitor impedance plot may be necessary. These equations only really give boundary conditions for capacitor selection and do not necessarily compute the optimum value.

A less rigorous and less laborious method is to use a value selected from the impedance analysis curves (doing the bypass calculation for a few PCBs is not too problematic, but for hundreds of ICs calculating the decoupling capacitance gets a bit tedious). Capacitors which are still capacitive and have a low ESR at the maximum switching frequency will provide suitable decoupling. Even capacitors which are beginning to behave inductively may be suitable providing their impedance is adequately low, as a general guide an impedance below 1Ω is adequate decoupling for most digital ICs.

Using the absolute switching frequency may not be adequate if there are very fast rise (τ_r) and fall (τ_f) times involved as these tend to produce harmonics that can dominate the frequency spectra. An estimate of the frequency content of a fast pulse edge (f_{edge}) can be obtained from the equation:

$$f_{edge} = \frac{1}{\pi\tau_r} \text{ or } f_{edge} = \frac{1}{\pi\tau_f} \qquad\qquad 2.13$$

Note: Usually the rise time is the dominant edge in EMC design and fall times are generally ignored. This is primarily due to the predominance of rising edge triggered logic.

For an ALS logic gate for example, even when driven by a 4 MHz clock, the rise time is faster than 4 ns, which gives a harmonic content of almost 80 MHz. Some may argue that significant harmonic content can still be present at up to 10 times this figure (see section on integrated circuits for suggested decoupling capacitor values for use with specific logic families).

The exact choice of capacitor for decoupling is difficult to predict accurately as many different effects may come in to play. If the rise and fall times are 10 ns or less, a value chosen on the clock frequency alone is usually adequate. With rise and fall times under 10 ns use the above equation to determine the upper operating frequency range of the decoupling capacitor and select a suitable value that meets this frequency specification.

Some recent experimental studies even suggest that values between 4.7 nF and 100 nF will be suitable for virtually all IC applications with clock frequencies above 33 MHz. The study also suggested that a 10 nF decoupling capacitor is an optimum value for digital circuits up to 100 MHz. Certainly this saves a lot of time in calculating suitable values and saves cost if all decoupling capacitors are of an equal value.

One thing is important to note, lead and track length connecting decoupling capacitors is even more critical than with bypass capacitors. The decoupling capacitor must have a minimal parasitic inductance if it is to operate effectively. Often the values of capacitance are already low due to the frequency range required, hence to have any decoupling effect at all lead and track length must be minimised (this is covered in the PCB section).

2.3.6 Parallel Resonance

When different values of capacitor, or different dielectric types, are connected in parallel (as with a bypass and supply decoupling capacitor) then a resonant mode can be introduced. Connecting capacitors in parallel is not necessarily a bad idea as it extends the operating frequency range of the capacitive effect (bypass or decoupling) as well as maintaining a low impedance to a large range of frequencies. However,

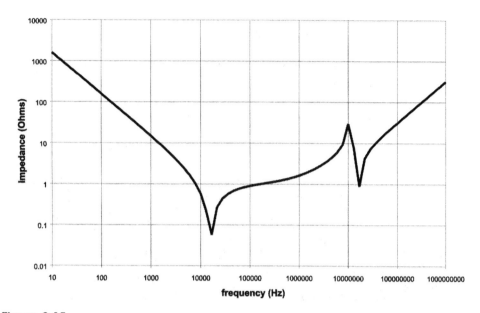

Figure 2.15

Capacitor parallel resonance

due to phase interaction and multiple parasitic inductances between each capacitor self-resonant frequency, there is a third resonance mode introduced.

These resonant modes will be present regardless of capacitor choice, the designer has little they can do other than to be aware of their presence. Because of this effect it is usually recommended that only one decoupling capacitor and one bypass capacitor be connected directly in parallel. If more filtering than can be achieved by two capacitors in parallel is required it is recommended that a series inductor or ferrite bead is used to separate sections of capacitors to increase filtering and reduce the interaction of these resonance peaks.

2.3.7 EMC Specific Capacitors

There is a special construction of capacitor called a feedthrough capacitor which exhibits very flat impedance-frequency response once the ESR value is reached. This type is used for filtering of signals into systems from external sources and can have a metallised body contact for direct mounting on to grounded panels or equipment bulk heads. They have been designed specifically with EMC in mind for signal and power line filtering applications, consequently they are expensive and offer only a limited range of values. There are ranges of SM feedthrough capacitors emerging in the market for low level signal lines, but limited values are available and their effectiveness relies on low ground impedance, hence proper PCB layout considerations for signal filter grounding is required.

In bypass applications there is a range of higher frequency aluminium electrolytic types which include an organic semiconductor in their dielectric. One type, sold under the trade name OS-CON™ (a trade mark of Sanyo Electric Company Ltd), offers extremely low ESR and high self-resonant frequency at high capacitance values (for example, a 47 μF, 16 V type has ESR <0.1 and self-resonant frequency >1 MHz). Certainly for circuits which contain high frequency switching power devices these types offer a better bypass solution compared with standard aluminium and tantalum capacitors of a similar value. As might be expected they are more expensive and are not available from as many sources as standard electrolytic capacitors. Similar aluminium electrolytic types with special polymer dielectrics are also being developed by other capacitor manufacturers.

For decoupling applications in combinational logic and memory circuits there are a range of flat plate capacitors which sell under the trade name MICRO/Q™ (a trade mark of Circuit Components Incorporated). These sit beneath various sizes of dual-in-line (DIL) and pin grid array (PGA) through hole ICs and some plastic leadless chip carrier (PLCC) SM packages. The parts are constructed from ceramic dielectric and fit into the same PCB holes as the IC and can be retrofitted to existing designs. These devices exhibit very low parasitic elements and occupy no additional PCB space, hence are extremely neat solutions to the decoupling of ICs, especially on PCBs which do not have a ground plane. The down side is that they are considerably more expensive than a single discrete capacitor, the supply and ground pins have to be in a specific configuration (i.e. 7 and 14 on a 14-pin DIL IC) and the range of values is limited. Despite the limitations the devices provide an extremely convenient solution to existing designs which were not correctly decoupled and new designs where there is no space for a discrete decoupling capacitor.

Figure 2.16

MICRO/Q decoupling capacitor for standard DIL logic gates, courtesy of Circuit Components Inc

The simplest way to avoid problems of self-resonance in standard capacitor types is to use components specifically designed for high frequencies (UHF, VHF and microwave applications, often on porcelain ceramic). This type of capacitor, commonly referred to as high-K, are more expensive than standard ceramic capacitors and may only be available in SM packages. The range of capacitance values are limited (up to 1000 pF) but handling and availability are comparable with standard SM ceramic types (MLCCs).

Three terminal capacitors are an attempt to accept that a lead has inductance and use a third lead to compensate by forming a T-filter. The component gives significantly improved high frequency performance compared with an equivalent valued 2 terminal type, but lead lengths should still be minimised, especially the ground lead connection. The frequency performance can be made to exceed SM ceramic capacitors if the ground lead is sufficiently short and connected directly to a low impedance ground (i.e. a ground plane).

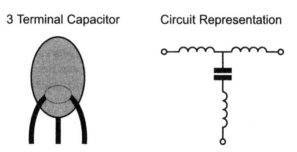

3 Terminal Capacitor   Circuit Representation

Figure 2.17

3 terminal capacitor

The EMC regulations have probably given one of the biggest marketing boosts in capacitor technology development over the last 20 years. This is likely to continue for some time with other dielectric materials, such as relaxor dielectrics, making a niche for themselves for certain EMC-specific problems.

2.3.8 EMC Critical Capacitor Applications

The most critical EMC application for capacitors are the bypass and decoupling applications already covered. There are further uses in filters and suppression circuits which also rely on capacitor performance to achieve the required circuit function.

In filtering applications it should be remembered that the capacitor has a limited frequency range over which it behaves capacitively. Low impedance is not a suitable parameter to select filter capacitors on as after self-resonance the phase angle is rotated 180° from −90° to +90°. If producing a filter choose a capacitor that is still acting as would be expected (i.e. capacitively). The best indications are a high self-resonant frequency and a low equivalent series resistance value. Also use the dielectric frequency table to ensure you have an appropriate choice of material. Much of a capacitor's ability to operate is lost when a long or thin PCB trace is used to connect it to its target circuit (this places an inductance in series with the capacitor terminal), the capacitor must be located close to its target circuit and connected with short traces.

In digital circuit decoupling applications it is a low ESR value that is more important than the capacitor's self-resonant frequency. Capacitors that are beginning to behave inductively, but still offer a low impedance path to ground can still provide adequate decoupling for most digital gates. This argument cannot be applied to analogue circuits as phase interactions can usually affect the circuit function.

2.4 Inductors

Inductors are a circuit element that a lot of designers prefer to avoid, probably due to a lack of understanding of their application and uses. Like capacitors, inductors are a potential cure for many EMC-related circuit problems and it is expected that their use will increase. Consequently, it is well worth spending some time re-evaluating them as circuit elements.

One of the reasons designers may have avoided inductive components is their potential to cause EMI as well as suppress it. The circuit forms a link between magnetic and electric fields, hence are potentially more susceptible than other components as they have an inherent ability to interact with magnetic fields. This conception may be based more on supposition than measured fact, even so it is worth considering the form or shape of the inductor construction prior to use as certain shapes are more problematic than others.

2.4.1 Construction

There are numerous shapes for the core materials used in inductor construction, but these can be reduced to just two types: closed loop and open loop. Open loop types are those in which the magnetic field deliberately passes through air to complete the magnetic circuit. The easiest open loop shape to imagine is the rod core (solenoid), in which there is a large magnetic field from the ends of the rod passes through the air around the rod to complete the magnetic circuit loop. Closed loop shapes can be considered as those in which the magnetic field is contained in the core itself (or is intended to be contained). The easiest closed loop to visualise is the toroid or ring core, in which the magnetic field is contained solely within the form factor of the core material.

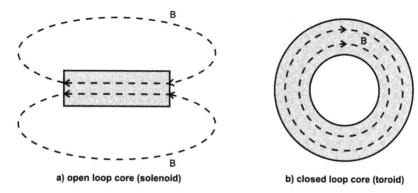

a) open loop core (solenoid)　　　　b) closed loop core (toroid)

Figure 2.18

Magnetic field in inductor core

Leads do not tend to be a problem on inductors as the parasitic inductance value is usually very small compared with the design value. Consequently, unlike most other components, the use of SM or leaded makes little difference to the EMC performance; this is dominated primarily by the core shape and winding styles. It is worth noting that SM inductors are only capable of supporting relatively low current values (up to a few hundred mA in most cases). Surface mount types are primarily aimed at signal lines where low current handling is not a problem. In power line applications the physical size and weight of the inductor core makes through hole the preferred mounting option.

The open loop inductor shapes are obviously the biggest potential EMC problem. The fact that the magnetic field deliberately exceeds the component body means that the potential for radiating a magnetic field is higher, and susceptibility to radiated fields is also increased. The coupling to other components positioned close to these open loop inductors is potentially high and they can be difficult to place in a circuit where high frequencies may be present or in which component density is high. The rod inductor shape is definitely the worse for EMC performance, the magnetic field fringes a large area around the rod core. The bobbin shape, which is also an open loop form and probably more popular than the rod inductor, has a relatively localised magnetic field, fringing to the open section of the core shape in quite a small air loop.

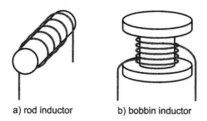

a) rod inductor　　　　b) bobbin inductor

Figure 2.19

Open loop conductors

Closed loop cores are often more expensive than open loop shapes and more expensive to wind, hence only a few closed loop core shapes are used for inductor components. More closed loop shapes are used in transformer design and will be discussed later. The most popular closed loop inductor is the toroid, which is also the least problematic as far as EMC is concerned. The toroid not only contains the generated magnetic field within the core itself, but any incident field radiating on to the shape creates an equal and opposite field in the winding, hence has a self-cancelling effect. The one other popular closed loop inductor shape is the pot core. This type has a closed loop surface, but requires exit slots for the wires. Consequently, some fringing around these slots can occur, but generally this is very small and highly localised. Pot cores are two part symmetrical cores and offer high inductance values and high operating frequencies, hence tend to be used in more specialised applications.

  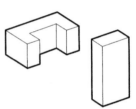

Figure 2.20

Some closed loop cores

To reduce cost of materials, inductors are often wound directly on to the core itself. The exception to this is the pot core where a bobbin former is used to hold the winding and enclosed in the core material at a later assembly stage. The winding sometimes dominates the cost of the finished part, hence the open core being less expensive as these shapes are simple to machine wind.

Unlike capacitors which have a large range of materials, inductors have basically only two choices of core material: iron or ferrite. This is a bit of a narrow classification as iron cored materials can include iron powder cores and laminated silicon–iron types. Likewise, there are a number of ferrite material compounds, the main two which account for in the region of 95% of ferrites are manganese–zinc (MgZn) and nickel–zinc (NiZn). The iron-cored devices are used for lower frequency applications (up to a few tens of kHz) as the iron compounds tend to have a limited magnetic frequency response. The ferrite compounds can be used into the high MHz frequency range, but are limited below a few kHz. Consequently, the use of inductors in EMC applications tends to be based on the ferrite cored devices.

There are also a range of inductors with no magnetic core: air-cored inductors. These are usually low cost, low inductance devices that operate over a wide frequency range consisting of a stiff piece of wire wound into a core shape (most often a rod). In general, they are not recommended for EMC applications as they exhibit extremely large flux leakage, offer very small inductance values (up to 1 μH typically) and act as receiving antennae to stray fields.

2.4.2 Inductor Impedance Analysis

Inductors, like capacitors, have a self-resonant frequency (f_o). If an inductor consisted of a single turn the self-resonant frequency would represent the ferromagnetic frequency limit (i.e. the limit that the material can 'flip' magnetic domains). In most inductors, with more than 1 turn, this self-resonant frequency is limited by the self-capacitance of the winding wire (C_w):

$$f_o = \frac{1}{2\pi\sqrt{LC_w}}$$

2.14

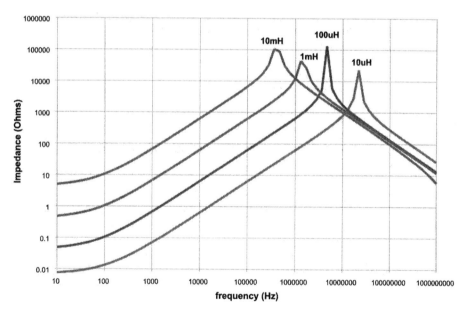

Impedance Analysis of Bobbin Inductors

Figure 2.21

Impedance of various inductors

The impedance curve of an inductor is therefore similar to that of a capacitor inverted, that is the impedance rises from DC to a high impedance peak, then falls as the parasitic wire capacitance dominates the impedance curve. There is a lower limit to the impedance determined by the DC resistance of the wire (R_{dc}) and an upper limit to the peak determined by magnetic loss in the core, analogous to the ESR of a capacitor. The magnetic loss can be considered for analysis as a resistive element (R_p) in parallel with the inductor in the electrical circuit.

Fortunately (unlike most capacitor data sheets), the value of self-resonant frequency is usually quoted in the inductor specification. This allows not only the appropriate inductor to be chosen for the circuit application, but gives some idea of the likely resonant modes that may appear with a wide band signal applied (i.e. likely attenuation of EMI).

The core loss factor is not quoted directly but implied by a parameter called the quality factor (Q). The quality factor is a measure of the inductive reactance to magnetic loss in the ideal operating band of the inductor frequency range (i.e. at a frequency where the impedance is dominated by the inductive reactance):

$$Q = \frac{R_p}{2\pi f_o L} \qquad\qquad 2.15$$

This equation can also be utilised in a circuit where the resonant peak is too sharp and causes reflection or ringing problems. Rearranging the equation for the resistance (R_p) can give a suitable resistor value to place in parallel with the inductor to produce a lower Q circuit. (Note that the quality factor of a single inductor is analogous to the quality factor of an LC tank oscillator circuit.)

2.4.3 Bias Dependence

Inductors also have a bias dependency similar to a ceramic capacitor voltage bias, but current rather than voltage dependent. The inductance value falls under applied DC bias current as part of the core is magnetised and consequently unavailable to attenuate AC signals.

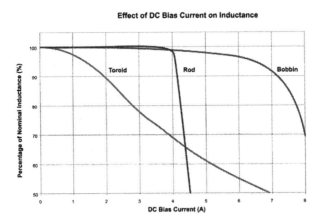

Figure 2.22

Inductor DC bias current

Not all inductors actually suffer this as much as others, high saturation cores (e.g. rod and gapped pot cores) tend to exhibit a thermal runaway effect before DC current saturation occurs. Toroid inductors tend to offer the lowest DC saturation for unit magnetic volume due to the enclosed core shape. Bobbin-shaped inductors suffer some saturation; however, the inductance value tends not to 'collapse' as rapidly as in toroid cores. Closed loop cores can be designed with the same values of current handling as open loop shapes; however, they are larger as they require a greater volume of core to achieve the same saturation characteristics.

There is therefore a hierarchy of selection on DC bias current which runs counter to the case for low emissions, as the lower emission cores (closed loop) suffer the most due to current saturation. The best way to avoid saturation when using closed loop cores is to ensure that the DC bias current is below the maximum value specified for the inductor. This current specification does not include any AC value, if there is a known significant AC current (often the AC value is not known in a noise suppression choke) the r.m.s. value plus any additional DC bias should not exceed the maximum DC current specification (I_{dc}).

The effect of this current bias-related droop of inductance is a reduction in the attenuation by the inductor on a noise or signal source. In an EMC-related application, such as a filter, this results in a reduction in the rejection and change in the effective frequency range. Usually, the changes are relatively small (less than 10% of nominal value) and a filter should always be designed to encompass the possibility of change in the passive elements under different load conditions. The effect is quite easy to simulate in a circuit simulation program (e.g. SPICE) or even in a spreadsheet.

2.4.5 EMC Specific Inductors

Ferrite beads are actually single turn inductors. The bead is usually slipped over a wire or has a single lead through the ferrite material to form the one turn. The advantage of single turns inductors is very high frequency attenuation range (the ferrite material is specifically formulated for high frequency operation) with very low losses at DC and low frequencies (up to several hundred kHz). Ferrite beads can also offer advantages in that they are easily retrofitted to a wire after the system has been designed. Hence if a high frequency problem is occurring that has only come to light at the end of a product design cycle these parts are easily added. The down side is that they have very low attenuation (typically 10 dB) and have to have a wire or placement available in the circuit.

Ferrite clamps act in a similar manner to ferrite beads, except they are usually applied over a cable bunch or ribbon cable. They offer a degree of both common mode (CM) and differential mode attenuation. Attenuation again occurs in the high MHz region and the clamp can be retrofitted to an existing design or cable installation. As with ferrite beads only a low level of attenuation is obtained (10–20 dB). Retrofitting is simple, most cable forms can have a clamp placed over them which has been specifically manufactured for the cable form. The effect this

Figure 2.23

Ferrite clamps, courtesy of Richco International

Ferrite Bead Conducted Attenuation

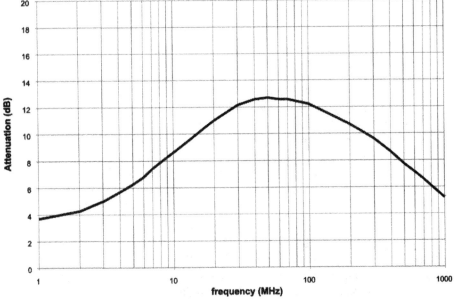

Figure 2.24

Typical attenuation of ferrite bead

could have on signal skew should be borne in mind if there are high frequency signals in the cable, especially single-ended signals.

Both beads and clamps are especially useful in dissipating transient energy from electrostatic discharges induced into a wire or cable loom. The ferrite dissipates much of the fast rising edge energy and dampens the transient seen at the exit of the wire or cable. More attenuation can be achieved by using larger ferrite beads and clamps or by adding more to either the same wire/cable or to additional wires and cables. These components do not necessarily have to be designed in at the start, but several SM bead components which include the single wire are coming on to the market, suggesting that these products are not used solely as retrofit components.

2.4.6 EMC Critical Inductor Applications

Inductors are commonly used for energy storage in DC–DC converters where the inductor form can have a significant effect on the EMC performance of the circuit. In power circuits the preference would be for toroid cores due to near zero emissions;

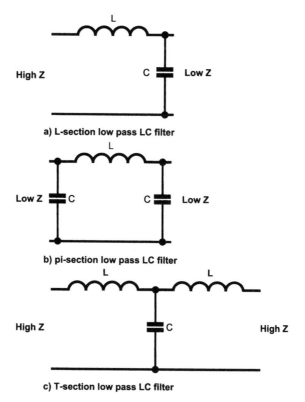

a) L-section low pass LC filter

b) pi-section low pass LC filter

c) T-section low pass LC filter

Figure 2.25

LC filter sections

however, if the toroid saturates high current conducted noise could be introduced. Often the bobbin shape is used as a good compromise between low emissions and high saturation current.

The best use for inductors is in the filtering of power supply lines where the low DC resistance produces very low losses. In particular, LC filters can be easily arranged to form an impedance matching bridge between different impedance circuits, such as a low impedance supply and a high impedance digital circuit. High impedances should be connected via a series inductor whereas low impedance circuits can have a bypass capacitor in parallel. The simple passive filter is easy to design and implement and high order frequency rejection can be easily obtained by cascading filter sections.

2.5 Transformers

The application area for a transformer usually decides the selection criteria and the EMC benefits are often considered 'in-built' due to the offer of galvanic isolation. There are areas which can help improve the EMC performance but often these can be at the expense of functional performance.

The transformer as a passive component is often overlooked in the same manner as inductors. In mains isolation applications often the lowest cost and smallest power units are used, in switched mode power supplies size of component dominates and usually means the higher frequency the better, and in signal isolation pulse shape determines choice. In all these application areas there are other considerations which can be used to ensure EMI is minimised.

2.5.1 Construction

The construction of mains isolation transformers with laminated iron cores for linear supplies is limited to virtually two choices: E cored or toroidal. The E core is the most popular due to lower cost and small size. Toroidal cored mains transformers cost more in both materials and winding cost and are usually significantly heavier as well as larger than an E cored transformer of the same power rating. As with inductors, however, the toroid offers advantages for reduced emissions and susceptibility even at low frequencies.

Switched mode power supplies (SMPS) usually utilise a ferrite core material and the most popular core form factors are pot cores, RM cores and E cores (including planar E core for very high frequencies). The toroidal core is unfortunately not as popular in SMPS due to its saturation capability (see section on inductors); hence from an EMC viewpoint the pot core and RM form offer the best shapes for reduced radiated emissions and susceptibility.

Signalling transformer construction will depend on the frequency of signal to be transmitted. Audio and telecommunication interface transformers are usually similar to mains linear transformers, that is an E core of laminated iron construction. Similar arguments can be applied as for the mains transformer. Low kHz signalling

Figure 2.26

Toroidal mains transformers, courtesy of Newport Components Ltd

transformers and pulse transformers require shapes offering high pulse retention but in small sizes, hence CI cores, pot cores and EP shapes dominate this frequency spectrum. Higher frequency signals (100 kHz to MHz) are commonly wound on toroidal ferrites, offering the best EMC performance of all signalling types.

2.5.2 Power Supply Transformers

Mains isolation transformers are generally designed to operate at low frequencies (50 or 60 Hz) and the use of laminated iron cores means they are not particularly good at conducting high frequencies (the 3 dB point of most mains transformers is typically about 1 kHz). However, at higher frequencies it is capacitive coupling not magnetic coupling that links circuits on either side of the isolation barrier. The parameter which gives some indication of the amount of conducted noise that may be transmitted is the isolation or interwinding capacitance. This should be low for linear mains transformers as the physical separation requirement for safety regulations can result in only a few hundreds or a few tens of picofarads of interwinding capacitance, resulting in very good high frequency noise rejection up to several MHz. With in-line filters and decoupling capacitors usually attached to rectification circuits the linear power supply can offer a very good conducted noise barrier for high frequencies, often sufficient in themselves to require no further noise filtering if power demand is low (under 50 W). Higher power linear supplies can usually be improved by the addition of simple line filtering using appropriately rated inductors and capacitors.

Figure 2.27

Laminated EE cored mains transformers, courtesy of Newport Components Ltd

Owing to the nature of SMPS operating at high frequencies (typically 25 kHz and higher) the circuit produces much more conducted noise than linear supplies. The SMPS has gained popularity due to high efficiency and small size (about 10 times reduction in size compared with similar power linear designs) and not its EMC friendliness. Additional filtering will be required if using an SMPS circuit as a direct off-line supply. Even for in-board DC–DC conversion SMPS circuits produce a high conducted noise and require filter considerations (see section on DC–DC converters). The filtering required to reduce the noise from the SMPS circuit itself is usually a good filter for other incident noise generated by the target circuit or injected into the input line. However, as the filter is usually not specifically constructed for frequencies much above the SMPS oscillator frequency, additional line filtering and protection may be required. Another potential drawback with SMPS circuits is that the interwinding capacitance of the transformer is often quite high to reduce stresses on the power components. This can result in SMPS transformers providing a lower level of immunity to conducted noise than a linear transformer.

2.5.3 Pulse and Signalling Transformers

At low frequencies up to the audio band (<20 kHz) signal transformers tend to be quite bulky as signal hold-up requires significant energy storage. For telecom and some audio circuits the silicon–iron EE core construction dominates. These offer a good high frequency noise rejection until the coupling capacitance begin to be the dominating impedance and high frequency signals can capacitively couple over the isolation barrier. Unless the attached circuit already has some high frequency rejection (i.e. slow amplifier circuits) then some additional simple filtering of the

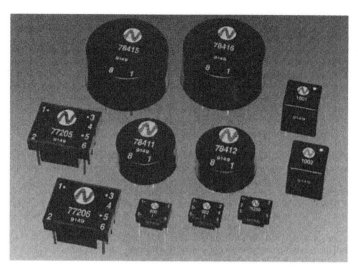

Figure 2.28

Pulse transformers, courtesy of Newport Components Ltd

signal, such as an RC or diode snubber, may be required. Another simple method is to use a similar transformer with a 1:1 ratio as a CM choke.

The best low kHz frequency construction for radiated emissions are those with almost closed outer surface, such as EP cores, as there is little or no fringing along these surfaces. Other shapes that can work well are those which approach the configuration of a toroid (e.g. C-I cores) and attempt to maintain the field within a predefined loop structure.

At higher signalling rates (20 kHz up to several hundred kHz) toroidal transformers are common and radiated EMI is close to zero. Conducted noise is usually controlled by matching between signalling transformers and careful selection of cabling and stub locations (e.g. Ethernet). Separated windings offering low interwinding capacitance can be easily implemented in high frequency signalling parts, improving the noise immunity further.

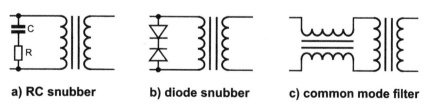

a) RC snubber   **b) diode snubber**   **c) common mode filter**

Figure 2.29

Transformer filter circuits

Layered windings help maintain reduced coupling capacitance and internal physical separation also helps. Some damping to minimise overshoot can also be applied either resistively or with a diode snubber type circuit external to the transformer.

Additional care to reduce the effect of reflections and potential ringing within signalling systems can be achieved by careful selection of the impedance profile of the signalling transformers. The transformers are optimised to have a peak impedance across their network or signal side at the optimum transmission rate. Consequently, the impedance at the signalling rate is higher than the out of band impedance. The peak impedance can, however, be too high and present a sharp impedance boundary which can cause overshoot in the signal and hence create a high frequency harmonic. The in–band impedance should be between 10 and 100 times that of the just out–of–band impedance, but less than 1000 times. This more gentle roll off of impedance produces a small benefit to the overall EMC of a single system, but as signalling can affect several systems is often worth the additional consideration (see section on transformer impedance analysis).

2.5.4 Transformer Shielding

Interwinding capacitance (sometimes referred to as isolation or coupling capacitance) of transformers can be significantly reduced by the use of shielding between the windings. An interwinding shield, termed a Faraday shield or electrostatic screen, consists of a thin conductive sheet layered over the windings. This can be connected to a common point on either primary or secondary winding, usually the local ground or the supply is adequate. If the ground is not a reference point at the transformer (e.g. a switched ground circuit) the shield should be connected to the steady-state terminal and not the switched terminal.

Additional shielding can be added for primary and secondary local grounds with an intermediate safety or mains ground between them. This level of isolation shielding

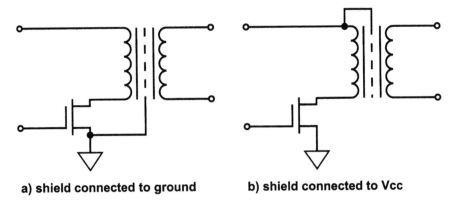

a) shield connected to ground   **b) shield connected to Vcc**

Figure 2.30

Transformer shield connection

can produce interwinding capacitance in the sub-pF range but adds to the expense of the transformer.

In low frequency and power applications the wave shape is usually unaffected by shielding but in signal applications a reduction in rise time could result if shielding is applied to high frequency signals (the shield provides capacitive coupling to ground, hence acts as a low pass filter). Also in high power SMPS applications this reduction in interwinding capacitance is often associated with a rise in leakage inductance which can stress the power components.

Completely enclosed shielded transformers are available, particularly for mains applications. The actual benefit of the shield is mainly in preventing low frequency mains hum from interfering with components located close to the transformer. The shield may give the winding wire some additional immunity to RF interference but this is not the original intention of the shield and the benefits are small.

2.5.5 Transformer Impedance Analysis

As with inductors, transformers have a self-resonant frequency. This is not often quoted for mains and low frequency transformers, but for signalling transformers this is usually close to the peak signalling frequency. The peak impedance shape can be used cautiously as another indicator of EMC performance. A very sharp peak impedance gives very high rejection to 'out of band' signals but can produce an easily resonant circuit with cabling capacitance, particularly with signal edge harmonics. A smoother peak curve can offer good rejection while reducing the possibility of edge harmonics resonating and creating a false signal.

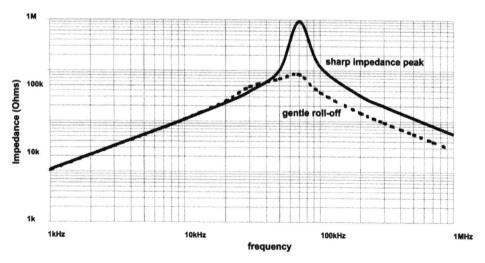

Figure 2.31

Network transformer impedance analysis

The impedance analysis of a transformer can be a useful plot to have for selecting impedance matching transformers where signal isolation over cables is required. Additional series resistors may be required, in series with the cable, if insufficient characteristic impedance at the transmission frequency is not available at the terminating end. Incorrect termination impedance causes reflections as well as signal attenuation. The greatest potential threat from a reflection are false data, emission problems would be stronger from the original signal source.

2.5.5 Common Mode Chokes

CM chokes are usually treated as inductors as their primary objective is filtering, but their construction is identical to a 1:1 transformer. CM chokes are most commonly wound on a toroidal core to minimise emission and susceptibility problems as they may be placed in close proximity to other filter components and at signal input/output ports.

This type of device is designed specifically for noise reduction and filtering, the important parameters are primary and leakage inductance. The primary inductance provides differential mode filtering, similar to a single inductor on the line. The leakage inductance produces a filter for CM noise (i.e. common to both lines) between each line and the system ground if capacitively coupled, or simply magnetic absorption of CM noise between the two lines when used without reference capacitors. The level of filtering will depend on the additional capacitors added or the parasitic capacitances within the system.

The easiest way to consider the action of a CM choke is in a filter application, such as a mains filter. Examination of a mains filter provides a good background to any passive filter design. The additional considerations in mains filter design is the earth leakage current requirement. This is a factor that should be borne in mind at each stage of design when using capacitors to provide an RF sink to a safety ground. If the safety ground or chassis is common with the mains earth then leakages must be summed to ensure the safety leakage current requirements are still met.

A standard mains line filter usually consists of five components; an X class capacitor at the inlet between live and neutral (CX1), a CM choke (sometimes referred to as a current compensated choke) between inlet and outlet (L1), an X class capacitor across the outlet sockets live and neutral (CX2) and finally two Y class capacitors between the outlet sockets and earth (CY1, CY2). Note that the mains supply is connected at the inlet side and the equipment at the outlet side. In most systems it is also advisable to add a high value discharge resistor (R1) to ensure that the capacitor charge is dissipated once power is removed, this prevents possible shock risk at the exposed mains pins. In off-the-shelf mains filter units the discharge resistor is often omitted as the user can fit this, or if feeding directly to a linear transformer this resistor is not necessary. Also shown are metal oxide varistors for protection against line transients; again these are optional and rarely included in standard filter units.

The mains filter operates in two filtering modes, differential (between the AC supply lines) and CM (between either AC supply and earth). The differential mode of operation filters out the noise between the supply lines reaching the mains and

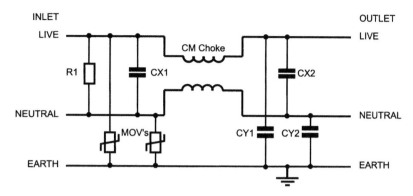

Figure 2.32

Mains filter circuit

likewise noise on the mains reaching the system. The CM filtering reduces noise from ground loops and earth offsets between systems.

The equivalent circuit for the differential mode signal is shown as a standard pi-filter. The X class capacitors dominate this circuit and often, where these have significantly greater capacitance than the Y class capacitors, the Y terms can be ignored. The effective inductance in each line is the leakage inductance (L_L) of the CM choke, and consequently this is much lower than the value usually quoted for the single inductance of one winding (for example in a typical 1.6 mH mains choke, the leakage inductance is only 30 µH). The leakage inductance is defined more by the constructional method for mains CM chokes, due to the safety requirement for separating live and neutral there is often little choice between types and a low value of leakage inductance is to be expected. In signal line filters, where the safety requirement is not critical, bifilar wound CM chokes can be used; however, leakage inductance will still be low compared with the primary inductance value.

Choice of capacitor value can be limited, typically X class capacitors are available from 1 nF to 2.2 µF, in mains applications values between 0.1 and 0.47 µF are common place. The 2 X class capacitors can be of different values, in fact this can

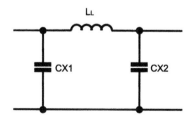

Figure 2.33

Differential mode equivalent circuit

be beneficial as it is unlikely that the impedance of the input mains line and system are equal. This does make frequency analysis of the filter more difficult without recourse to simulation, but allows impedance matching at both inlet and outlet terminals. For high frequency signal filtering lower value capacitors would be used.

As far as the standard EMC conducted tests are concerned we know that the differential mode impedance between live and neutral is 100Ω (CISPR 16 $50\Omega/50$ μH LISN). If the filter is supplying a switched mode power supply, it is likely to have an impedance in the 0.1Ω region, hence matched X class capacitors are not desirable and removing CX2 completely may be beneficial. Most off-the-shelf filters are designed for 50Ω circuit impedances, and hence may not be the ideal for many situations.

The cut-off frequency is not as easy to estimate due to the inductors in both branches; however, with equal value X class capacitors and negligible Y class (CX >50 CY) then an approximation can be derived from:

$$f_c = \frac{1}{2\pi\sqrt{L_L C_X}}$$

2.16

The 2 X class capacitors (and any Y class capacitance) produce the pi-filter arrangement, giving a theoretical fall off of 60 dB/decade after the cut-off frequency (in practice 50 dB/decade is more typical).

The CM equivalent circuit is a simple LC low pass filter, present in both the live and neutral lines. The inductance available is the full inductance of each single arm of the CM choke, and the combination of Y class capacitor completes the circuit. The cut-off frequency for CM signals is:

$$f_c = \frac{1}{2\pi\sqrt{L_p 2C_Y}}$$

2.17

after which the attenuation is close to 40 dB/decade based on 50Ω terminating impedances.

CM chokes for signal lines can be analysed and designed into a circuit in the same way as the above mains filter. DC current saturation is usually much higher than with single inductors as the DC magnetic field is self-cancelling, hence a high saturation current can be obtained on a toroid core.

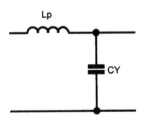

Figure 2.34

Common mode equivalent circuit

CHAPTER 3

ACTIVE DISCRETE COMPONENTS

There are almost as many applications for active discrete semiconductors as there are for the passive component; however, the selection of device type for specific applications is usually more stringent than with passive devices. Consequently, the choice of final component may not be influenced as much by its potential electromagnetic compatibility (EMC) effect as can be applied with passive devices, but probably more with respect to its functional performance in the application.

This sounds like an excuse to ignore or gloss over EMC issues within discrete semiconductor devices. This is not the case, but it should be borne in mind that some of the suggestions and design tips quoted here may ultimately not be as applicable to the circuit as the suggestions for other components, due to the functionality of the final circuit. The EMC issue should never stop a circuit design for a specific application being produced and it is not intended to either specifically increase design complexity or eliminate certain circuits from the world. The suggestions contained here are just that, suggestions. At the end of the day it is the circuit functionality in its end application that is paramount; EMC at the design stage should only guide a designer, not completely restrict their output, and should hopefully have a minimal influence on overall circuit costs.

3.1 Discrete Component Packaging

The number of types of packages available for discrete semiconductor devices probably exceeds that for integrated circuits (ICs). The packages tend to have between two and four contacts (leads) for the component itself, sometimes with additional pins for multiple components in the same package or for heat dissipation of the device. Either way the devices have few contact points and the device of interest may be available in a choice of package styles.

The assembly of a discrete device into a package is similar to an IC, the device is usually bonded by epoxy die attach or eutectic metal to the leadframe of the package. Bond wires are used to connect the device to the other package leads; hence, unlike passive devices, many of the device terminals have a 'wire' inside the package. The bond wire length and dimensions are similar between package and component types, typically having in the region of 0.5 nH of inductance, this is usually less than the lead inductance itself; hence lead inductance is the dominating parasitic element.

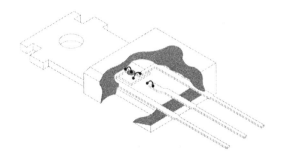

Figure 3.1

TO220 packaging of discrete semiconductor

As with virtually all device packaging with respect to EMC, surface mount is preferred. If the parasitic associated with a small transistor or diode pack are compared, using SOT23 as the surface mount pack and E-line (similar to TO92) for the leaded package, it is observed that even with minimum lead length on the through hole package it still has twice the parasitic inductance and capacitance as the surface mount package (Table 3.1).

Table 3.1 *Comparison of surface mount (SOT23) and through hole (E-line) package parasitics*

Package type	Lead and bond wire inductance (nH)	Inter-lead capacitance (pF)
SOT23	2.5	0.08
E-line	5	0.15

Not all package styles have such a marked difference between surface mount and leaded types, but the surface mount versions always have lower parasitic elements for a similar size package as it minimises the lead lengths within the package design. Shorter lead lengths also minimise the magnitude of return loops and potential aerial sizes, but at this level of dimension the differences between surface mount and through hole are negligible compared with printed circuit board (PCB) track lengths contributing to radiated effects.

3.2 Diodes

Diodes are the simplest of active semiconductor devices, they consist essentially of two dissimilar materials, either two oppositely doped semiconductors (p- and n-type) or a metal and a semiconductor (Schottky diode). Other more exotic combinations do exist, but essentially the diode is as straightforward as a semiconductor can be.

Like inductors and capacitors, diodes are a potential cure to many EMC ills, especially electrostatic discharge (ESD) problems. There are several diode circuits and formulations of diodes, which are in fact specifically aimed at EMC solutions. Hence, although the simplest of semiconductor parts, diodes are probably the most relevant to EMC issues at the discrete component level.

A feature of diodes which is generally unique in terms of EMC, is that the faster components are generally better for the EMC performance. This is unusual as most of the time the designer is trying to limit bandwidth and reduce the frequency response of the circuit (outside of the operational frequency response) to reduce EMC problems. Diodes are possibly the only device in which a fast response time is beneficial to EMC, but again there is the caveat that this is only in certain applications and most of the time slower diodes that are perfectly adequate for their function will not have a noticeable effect on EMC.

The parameter which most effects a diode's response or speed of operating is its reverse recovery time (τ_r). This is the time taken for the potential barrier created under a forward bias (typically the potential barrier is 0.6 V for a p–n junction diode) to recombine and prevent reverse current flow. Until the barrier is recombined a reverse current can flow through the junction, hence potentially a reverse current spike can be observed with a pulse width equal to the reverse recovery time. The reverse recovery time is often in the nanosecond region, but varies with diode size (increasing with diode size and current carrying capability), construction (lower for low barrier types such as Schottky diodes) and bias conditions (lower recovery times at lower reverse bias, but this is only a marginal effect). Consequently, as a quick check on diode performance for EMC considerations, lower transit time devices (i.e. fast switching diodes) are preferred.

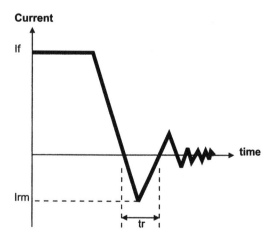

Figure 3.2
Diode reverse recovery

3.2.1 Small Signal Diodes

Diodes used in applications involving signal clipping or providing offset voltage levels can be classified as small signal diodes. The applications do not involve large current handling (typically under 1 A) and often the diode is only used under low signal voltage conditions. Owing to the lower current requirement the diodes are quite small and most commonly p–n junction diodes are used in these applications because of their low cost and wide availability.

As this type of diode is geometrically small the recovery times are quite low (about 10 ns would be typical), and hence usually react to all but the fastest transient signals. The standard small signal diode is therefore quite useful for a low-cost method of protecting circuits from low current discharge and small transients, such as DC line spikes and in-board ESD residues. It is not necessary to have this type of diode protection on all signal lines, only those which are accessible to external contacts, sockets or cables.

Often many ICs feature p–n junction diodes immediately beneath the bond pad on the silicon. This gives the IC a level of ESD protection, the lead frame of the package and bond wire add a small inductance to reduce the total discharge current into the IC circuit. Being beneath the bond pad means that the structure takes up no additional silicon area, hence it is a popular feature and usually available 'free' to both designer and user of the IC.

3.2.2 Rectifier Diodes

Some small signals are used to rectify low level AC; however, the main application of rectifier diodes are in linear power supply units (PSUs), rectifying transformed or direct mains AC, and in switched mode power supplies (SMPS). There are significant differences in these applications, in linear supplies the diodes deal with a slow AC signal (50 or 60 Hz) and generally are large slow diodes of p–n junction construction and relatively low cost. The rectifiers used in SMPS applications are

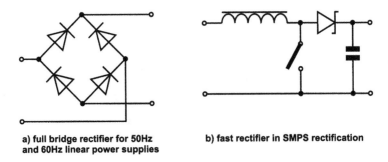

a) full bridge rectifier for 50Hz and 60Hz linear power supplies

b) fast rectifier in SMPS rectification

Figure 3.3

Rectifier diode circuits

usually fast action diodes as switching can be in the kHz to MHz frequency range, often of Schottky construction and relatively expensive.

With linear PSUs the diode is used solely as a rectifier and not for any ESD, transient or mains line protection purposes. If transient or ESD protection is required it has to be provided elsewhere as the rectifier diodes only connect between the supply lines and load, not usually to the mains earth. The rise times seen by the diodes from the mains to load may not be as slow as the mains frequency sinusoid due to the supply storage capacitors holding up the voltage; hence the diodes may experience relatively short bursts near the peak of the mains sinusoid. Even with these short bursts the rise times are relatively slow and, consequently, it is quite acceptable to use slow, low cost, bulk silicon diodes and EMC considerations are applied elsewhere in the circuit.

In SMPS circuits there are potentially fast transient spikes from the magnetics within the circuit, as well as any harmonics from the switching circuit pulses themselves. Consequently, fast-acting diodes are required in this application and the speed of response is critical to minimising the amount of noise generated, therefore energy wasted, by the SMPS circuit. By their very nature SMPS circuits are noisy, using fast switching waveforms to pulse a magnetic storage element. As well as the switching action with its characteristic switching frequency, the rise and fall times of the pulse generate harmonics. There may also be some ringing and overshoot in the magnetic circuit; hence there can be a large amount of high frequency harmonic content which is unrelated to the actual switching frequency of the circuit.

Fast rectifiers and Schottky diodes are most commonly used for the catch diode or feed diode in SMPS circuits (see next section for Schottky diode). Fast rectifiers are essentially p–n junction diodes with a construction which produces a fast transit time (i.e. a thin junction barrier within the diode) and are more popular in boost and feed-forward applications. The faster the diode the more of the harmonic content it will handle, hence the lower the potential conducted EMC of the completed circuit as higher frequency pulses are passed on to the storage capacitor and are not reflected by the diode. The capacitor has to be able to handle these high frequency signals for the fast diode to be useful and often a combination of capacitors and passive filtering is used.

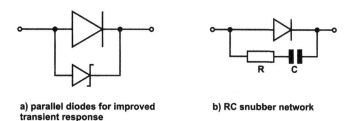

a) parallel diodes for improved transient response

b) RC snubber network

Figure 3.4

Improved transient circuits

At high current levels the diode junction area is going to prevent fast switching, the capacitance of a diode is proportional to its junction area, hence the high current diodes (> 1 A) are likely to be slow (τ_r > 15 ns). One solution to this is to have a smaller diode in parallel that has a higher reverse breakdown and higher forward voltage drop. Under normal conditions this diode is not in operation as the lower voltage drop bulk diode takes the current. On transients and fast edges this small diode can catch the lower current transient and dissipate this small amount of power before the slower diode operates on the bulk energy. This paralleling can be effective at removing the majority of fast transient spikes providing they have a low power content, but pairing diodes for this application can be difficult as the amount of power in the transient is not known.

A common diode noise suppression circuit is the RC snubber network. This uses a small capacitor as a DC block and a low value resistor to dissipate the transient energy. Low values of resistor and capacitor are selected so that the RC time constant is significantly lower than the transit time of the diode (see Equation 3.1).

3.2.3 Schottky Diodes

Schottky barrier diodes offer a very low forward voltage drop, high current density and fast reverse transit times compared with similar sized p–n junction diodes. This makes the Schottky diode a good protector against fast transient signals and spikes. The cost of a Schottky diode is generally higher than a p–n junction so they are used sparingly in circuit design.

One of the common uses of Schottky diodes is as an alternative to a fast rectifier diode in an SMPS circuit. All the arguments previously raised for a fast rectifier apply (i.e. fast reaction to both forward current and reverse transient, spike protection) with the additional benefit in SMPS circuits of a low forward voltage drop, hence low loss.

The EMC of a circuit using Schottky diodes should be slightly better than with junction diodes; however, the actual magnitude of the benefit may be small. It is unlikely that switching from p–n junction to Schottky barrier diodes will produce a cost-effective improvement in the EMC performance of most circuits. However, additional efficiency savings due to the lower forward voltage drop and consequent better voltage regulation characteristics should justify the cost in power circuits and especially in high frequency SMPS circuits.

3.2.4 Zener Diodes

Zener diodes are normally operated in the reverse mode, they feature a sharp (abrupt) reverse voltage current transport action. The Zener action reacts quickly to reverse voltage transients. Zener diodes are often used as protection against over voltage, hence are a good protection against ESD discharge providing they can tolerate the power level (i.e. current) of the discharge.

Zener diodes are used in signal lines for ESD protection as they offer a low cost alternative to specialist protection and can be used easily at the voltage levels found in the majority of signalling applications. In signal lines a Zener diode can offer a low capacitance compared with other solutions, hence do not cause significant signal distortion on high data rate signalling lines.

In applications where the Zener diode is being used in-board (i.e. not at the input circuits as protection devices) the operation of a Zener diode can act as a potential EMC noise problem. Owing to the fast nature of the Zener action, when this occurs a relatively large current surge can be observed as the Zener conducts the reverse over voltage current. Consequently, circuits using in-board Zener diodes for over voltage protection will require some localised decoupling for the Zener action. Another method to limit the magnitude of the surge is to use an in-line feed resistor to the Zener. If using a feed resistor it needs to be relatively low value to ensure that the action occurring is the Zener action and not another effect, such as avalanching within the junction.

Zener diodes with 12 V operation are particularly prone to becoming wideband noise generators if a feed resistor limits the current and the diode skips between Zener action and avalanche. This activity can be used to make a noise generating test circuit, but should be avoided in most applications.

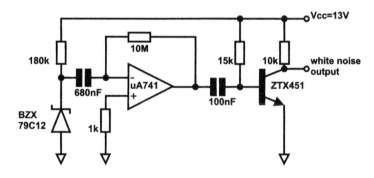

Figure 3.5

Noise-generating circuit using 12 V Zener diode

3.2.5 Light Emitting Diodes (LEDs)

In themselves LEDs are unlikely to be a problem with regards to circuit EMC. The device is most often operated under a DC bias in forward conduction mode and therefore has little direct effect on the circuit.

The main application problem that could occur is when the LED is mounted into a panel at a distance from the circuit whose operation it is an indicator for. If the leads to/from the LED act as an aerial they could pick up radiated emissions from other circuits or from external sources. Likewise, if the DC bias for the indicator has a noise from the indicated circuit superimposed on it the leads could act as a radiating aerial. These problems are strictly in the application of long leads and the LED itself is not an EMC threat.

3.2.6 EMC Specific Diodes

Transient voltage suppressor (TVS) or transorbs are a form of silicon diode that feature similar characteristics to Zener diodes, but act in an avalanche mode. One main difference is the absolute value of clamp voltage, a Zener will have a very tight specification, say 12.3 V±2%, whereas a TVS rated for 12 V lines may have an avalanche voltage between 13 and 19 V. TVS diodes in general can cope with higher current levels than a similarly sized Zener diode and can be configured to operate unidirectionally (i.e. have similar forward and reverse characteristics) and hence can cope with both positive and negative transients. These devices are usually specifically designed to cope with the very high voltage transients seen from ESD, lightning-induced transients and mains spikes (>500 V transients). The unidirectional device is particularly suited to signal lines which signal between a high value and ground (single-ended schemes), bi-directional devices are used where both positive and negative voltage signalling levels are used (differential signalling) or on AC lines. A feature of bi-directional devices which make them popular in many signalling applications are their very low junction capacitance (typically 5 pF) compared with a similar power rated unidirectional device (500 pF).

Varistors are a formulation of metal-coated ceramic pills pressed into a single assembly. Each ceramic pill works as a Schottky diode with a high potential barrier of approximately 3.6 V per boundary; hence voltage types are made up from layers of these pills. Production techniques have enabled the devices to be made in a controlled multilayer approach (multilayer varistor, MLV), similar to a ceramic capacitor; hence surface mount versions are quite popular providing a large barrier area (i.e. high current capability) in a small volume. The devices are also sold under various names including voltage-dependent ceramic resistors (VDR) and metal oxide varistors (MOV). It is more common to find this type of product being manufactured by ceramic capacitor suppliers than silicon diode vendors due to the manufacturing processes involved.

The varistor is a very good alternative to a Zener or TVS diode for line protection particularly on mains inlets. The main drawback of a varistor compared with diode clamping is that the value of clamping voltage tends not to be as well defined; hence the circuit needs to be able to cope with this clamp voltage tolerance. The varistor can be used on signal lines (e.g. RS232, EIA485, SCSI), although its higher capacitance tends to limit the speed of data or size of varistor. Compared with Zener diodes, varistors can be obtained in smaller package sizes for a similar current handling capability. Varistor reaction time is in the sub-nanosecond region for the smaller devices; hence it can offer the fastest responses to transients when compared with Zener and TVS diodes and would be the recommended choice for first-line ESD protection and on high voltage, high transient risk lines (e.g. entry/exit cabling, power lines, etc.).

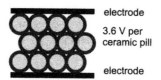

electrode

3.6 V per
ceramic pill

electrode

Figure 3.6

Varistor cross-section

3.2.7 Diode Applications for EMC Control

The majority of the above diode components have been discussed in relationship to their effect on transients, assuming that the transient is incident to the system (i.e. not generated within the circuit). There are of course instances where the circuit is operating inductive loads or switching large currents that cause transients to be generated within the system. Again the diode can be used at the source of the noise as a transient voltage suppressor.

Where diode noise is a problem a simple RC snubber filter across the diode should be adequate to reduce any switching transients from the diode reaching other circuits. A very low capacitor and resistor value are usually used, say 47 pF and 50Ω for τ_r >2 ns, depending on the diode noise pulse duration and rise time.

$$RC < \tau_r \qquad\qquad 3.1$$

Inductive loads are a particularly common source of transients and one that is usually easy to identify and deal with. Typically, relay energising circuits see a back EMF transient from the relay coil when the relay is de-energised; the energising current can also see an inductive overshoot creating a transient. These transients are best

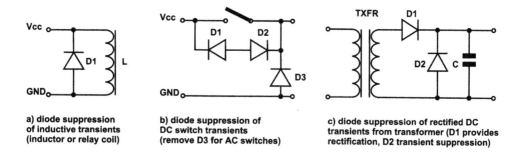

a) diode suppression
of inductive transients
(inductor or relay coil)

b) diode suppression of
DC switch transients
(remove D3 for AC switches)

c) diode suppression of rectified DC
transients from transformer (D1 provides
rectification, D2 transient suppression)

Figure 3.7

Diode suppression applications.

suppressed using clamping diodes across the relay coil. The diode should be positioned close to the relay even if the switching transistor is remote, the transient should be prevented from travelling in any conductor and hence radiating or coupling with other circuits.

In power transformer circuits the transients can be difficult to suppress without affecting the efficiency of the circuit. Some overvoltage protection can be included using relatively low-cost Zener or Schottky diodes to clip the maximum voltage excursion. This protects the output circuit from excessive output voltages without limiting the efficiency under load conditions.

Motor circuits are another inductive load that can benefit from diode suppression of commutator or brush noise. The exact configuration will depend on the motor type (DC, three-phase, etc.), but the suppressor should be located as close as possible to the motor contacts to reduce the risk of emissions from the connecting control or power cables.

Switch bounce can also generate a transient pulse into a system, these are often difficult to characterise and usually connected with high voltage switches rather than low voltage types. A varistor or Zener diode across the switch contacts usually provides a good level of protection.

In digital signal lines transients internal to the system are usually of relatively low energy, hence low cost Zener or low breakdown small signal diodes can be used. Where the signal exits the system then a higher level of protection would be recommended. In parallel data bus communications system the cost is going to be high if significant numbers of suppressor are required. It would be usual to use the lower cost varistors or small TVS diodes in this type of application.

On power line inlets the protection device has to be able to tolerate higher stresses than most board level components under normal operation. Consequently, high energy transorbs, TVS diodes or high voltage varistors are most commonly employed on inlet power lines. The power line could have transients from many sources, including other equipment connected to the same mains ring and even lightning-induced transients. The power line is one of the most useful places to place this type of component, it not only protects the equipment from incident transients, but suppresses transients from the system being conducted down the mains ring.

At signal connector interfaces to the outside world, ESD is probably the most common EMC problem to be encountered. This is especially true for connectors, which will have user contact due to their function (e.g. RS232 interfaces, printer ports, keyboard connectors, etc.). The cost of ESD protection can be reduced by protecting the contacts in a shielded connector with recessed pins and cabling with earthed screen. Without recessed pins in a shielded connector housing, varistor or TVS protection on the signal lines is likely to be required. Passive low pass filters could be used, but the advantage of active devices over passive filters are the low capacitance these add to the signal line, therefore very low distortion or skew is added to the signal.

3.3 Transistors

Transistors are traditionally classified as either bipolar or unipolar. The nomenclature describing the current carrier within the semiconductor, bipolar having both electrons and holes carrying current, unipolar having only one carrier particle. There has been a recent addition to the family of transistors that slightly clouds the simplicity of this; the insulated gate bipolar transistor. In the majority of cases when people talk about transistors they are describing the bipolar junction transistor, otherwise the unipolar devices are usually known by their acronym (e.g. FET, see below).

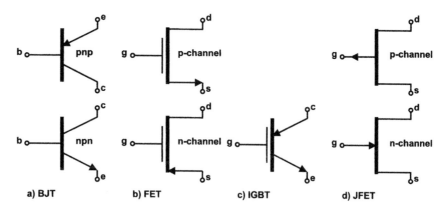

Figure 3.8

Transistor schematic symbols

Transistors are generally not employed for EMC performance but purely for their functional operation. Consequently, there is not as much EMC data that can be found on transistors as on diodes and applications where the transistor choice has affected the EMC are very difficult to find. However, there are still some component choices and options that may help with the EMC performance of the circuit if not affecting the individual components or function greatly.

3.3.1 Bipolar Junction Transistor (BJT)

A common way of describing a bipolar transistor in student textbooks is as two back to back diodes. This relatively simplistic description is a good enough analogy for examination of the bipolar transistors EMC characteristics. Because of the diode structure bipolar transistors are relatively insensitive to ESD and many of the diode arguments can be applied to the bipolar transistor. Consequently, bipolar transistors are a good choice for discrete input or output buffer circuitry which is accessible by the user, especially discrete amplifiers and user accessible analogue buffer ports.

Although relatively insensitive themselves, bipolar transistors should not be considered as protection against ESD. Usually the individual transistor junction diodes are too slow to react to transients on their terminals. The reference potential for the transistor may also not be the ground connection, and almost certainly won't be the safety ground. Consequently, it is probable that some other ESD protection will still be required if using bipolar transistors at an input or output, but at least the circuits containing bipolar transistors are less likely to suffer damage from ESD.

Smaller transistors are faster and consequently able to react to transient voltages better than larger bipolar transistors; however, having a smaller junction means the device is less able to cope with larger energies. Larger bipolar transistors can be slow compared with their small signal counterparts, hence as well as having good immunity to ESD damage, large transistors tend not to behave as transient or noise amplifiers. Some small signal bipolar transistors can be operated as switches or mixers in the GHz region, so unless the circuit requires this level of performance for its functional requirement, it is better to use lower cost slower bipolar transistors.

The main EMC threat to bipolar transistors is probably conducted noise. The base drive to a bipolar transistor works on a low voltage (typically 0.6 V) and the device is essentially a current amplifier. Consequently, a low amount of noise on the base would be amplified by the transistor. It is for this reason that the transistor should be selected with an appropriate switching speed, use devices with as low a switching speed as possible for the circuit function. It is also important to include any filtering at the base terminal as this is where the noise has its greatest effect.

Modulated radio-frequency (RF) signal fields can become demodulated by small signal bipolar transistors. The non-linearity of the voltage–current relationship (being exponential for both diodes and bipolar transistors) can allow a strong RF signal to be demodulated within the transistor. The RF itself is usually not conducted out of the transistor, but can build up a DC offset, possibly sufficient to cause forward conduction. The demodulated signal can fall within the operating frequency range of the circuit, hence can then become an interference signal within the circuit. The effect is most noticeable with demodulated audio signals, particularly with the proliferation of mobile telecommunications. The problem is not as acute in discrete circuits as in integrated circuits due to the larger dimensions of discrete transistor junctions and leads and interconnect between components acting as filters, but if using very small, fast transistors the potential for interference does exist and tests with a modulated RF source will be necessary to check circuit immunity.

3.3.2 Field Effect Transistor (FET)

Field effect transistors work on a gate charge rather than a base current as in bipolar transistors. This produces two effects as far as EMC performance is concerned. First, the gate is sensitive to charge rather than actual current flow so can be easily upset by a transients. As there is no conduction path from the gate transients or other noise can cause a charge build up. Secondly, at large gate junction areas the device can still operate at high speed; hence unlike bipolar devices where larger transistors represent reduced conducted EMC problems the FET is still a sensitive component, almost regardless of size.

A problem that was common with early FETs was latch-up, this is where a transient, usually on the drain or source connection, would charge the gate and latch the transistor permanently on. Most modern FET structures incorporate a body diode between the drain and source and a separate substrate connection (this is sometimes common with the drain). Modern FET components no longer suffer from latch up and are often used for switching inductive loads and lines containing transient spikes (e.g. relay coils, motors).

Transient protection of FET circuits should generally be applied to the gate. The gate is isolated from the drain and source connections by an oxide layer within the device, and hence does not have the benefit of the body diode protection. Usually, gate control circuits would be expected to incorporate this protection; however, if the FET is being operated remote from the control circuit (e.g. near its load, a motor for instance) then there may be a requirement for diode or varistor protection between the gate connection and a local safety ground. The leakage current of this additional component also reduces gate charge build up.

Gate bias voltage for a FET is usually in the high volt region (8 V for many power devices), although there are now popular 2.5 V devices for direct logic drive. The higher gate level can give the FET a higher immunity to conducted noise than a bipolar transistor. At the small feature size level FETs can also exhibit lower gate capacitance, this may be useful for the circuit functionality to achieve a high speed input at high impedance; however, on the EMC front this allows a small noise signal to achieve possibly enough charge to operate the gate. Unless the circuit requires the low gate capacitance and high impedance, an easy way to reduce the risk of noise-charge build up on a gate is to include an additional bleed resistor to ground on the gate, this should maintain the high switching speed characteristic.

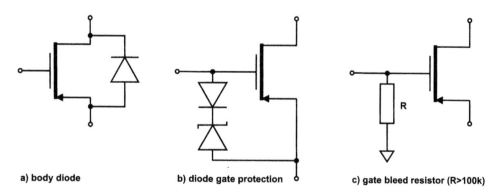

a) body diode   b) diode gate protection   c) gate bleed resistor (R>100k)

Figure 3.9

FET protection circuits

Modulated RF signals can also be demodulated by small FETs as well as small signal bipolar transistors. The non-linearity of the voltage–current relationship is a square law for unipolar devices, hence the demodulation characteristics are not as strong as in bipolar devices. The potential for charge build up due to RF creating a DC bias offset is possibly greater with FET devices due to the high impedance gate compared with the current drive requirement for a bipolar device. The effects of RF interference and bias offset are again more likely to manifest in an IC than a discrete circuit. High performance analogue circuits (i.e. high resolution, highly sensitive circuits) are much more susceptible to this type of interference than switching, power train, line drivers or digital circuits.

3.3.3 Insulated Gate Bipolar Transistors (IGBTs)

The insulated gate bipolar transistor is a combination of the insulated gate drive capability of the FET and the collector-emitter current handling of the bipolar transistor. The component was developed to achieve low loss switching at high speeds in power supply and motor applications, hence the construction. With regards to EMC little has been written but its analogy to the FET and bipolar transistor allow some estimation of its performance.

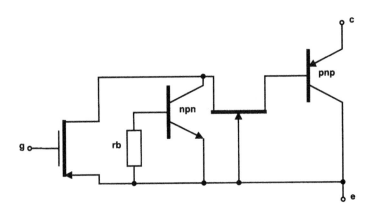

Figure 3.10
IGBT equivalent circuit

Having an insulated gate charge drive would suggest the device would be as sensitive to transients and ESD as a FET. Similar drive circuits are used for FET and IGBT devices, hence the same level of static and transient protection should be used. Having a bipolar structure across the other terminals does eliminate the need for a body diode and reduces the threat of transient interference from these terminals affecting the gate and the device may be slightly less sensitive than a standard FET, but it is unlikely to be significantly better. So like its structure, the IGBT has EMC characteristics between the FET and bipolar transistor.

Many IGBTs do not have a reverse diode optimised into the device across the emitter-collector terminals, in fact the reverse diode is a parasitic diode and a poor performance part. To reduce ringing and conduction losses when switching inductive loads either an external diode should be used across the collector-emitter terminals or an IGBT with a built in optimised diode should be sought.

3.3.4 Junction Field Effect Transistor (JFET)

The FET described in a previous section is strictly speaking a metal-oxide-semiconductor construction (MOSFET). The junction field effect transistor was a precursor to the MOSFET using a p–n barrier as the control element with the conduction usually occurring when no gate drive was present. These devices are not as popular in discrete form as the MOSFET but are still common on ICs where the structure is used to create high value resistors.

There is a recent circuit development in EMC protection that may increase the popularity of JFETs. The main reason for the reduction in popularity is the fact that the devices are 'on' when no gate control is present. This effect can be utilised in a circuit called a transient blocking unit (TBU) for protecting lines against transients by having the transient impulse operate the gate. The circuit requires some form of diode to ensure that the line is 'on' during normal operation and that the gate can only be operated by a rise at the input of greater than the gate turn-on voltage (V_{gs}) plus the diode forward voltage drop (V_f). The circuit as shown is a very simple implementation of the TBU, a bi-directional circuit can be made using three JFETs and four diodes.

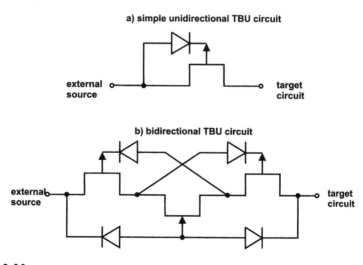

a) simple unidirectional TBU circuit

b) bidirectional TBU circuit

Figure 3.11

Transient blocking unit

The diodes need to be very fast to enable the gate reaction to cut off the transient before it reaches the target circuit. For power lines the devices will have to feature very low resistance in the 'on' state. The main benefit of this type of circuit is its position in series in the line means that it avoids dumping the energy to ground (possibly creating a ground bounce) and dissipates it in the body resistance. The circuit also can react proportionally to the incident energy. The availability of JFET structured resistors on ICs also means the circuit could be integrated into line driver/receiver ICs with minimum design or silicon overhead.

3.4 Other Active Discrete Devices

Diodes and transistors certainly account for the vast majority of discrete semiconductor devices used, however, there are other devices that sell in large numbers, sometimes for specific applications, and some are considered below.

3.4.1 Silicon-Controlled Rectifier (Thyristor)

The silicon-controlled rectifier (SCR) or thyristor is a semiconductor structure primarily used to control AC power. The construction is similar to two back-to-back bipolar transistors. Hence many of the notes for the bipolar transistor could be applied. Some critical differences are that the gate contact requires a current drive much higher than a bipolar transistor and usually operates at a much higher voltage, hence the device is relatively insensitive to conducted noise.

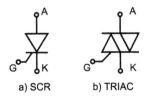

a) SCR   b) TRIAC

Figure 3.12

(a) SCR and (b) TRIAC schematic symbols

The bipolar structure and large junction area provides some degree of ESD and transient protection. The rate of gate rise usually has to be controlled to prevent device burn-out, hence transients of sufficient energy on the gate node could be a problem. These devices rarely suffer in use from EMC problems themselves and it is usually the drive circuitry that requires most consideration with regards to EMC and noise performance.

The main EMC concern would occur if using in a DC circuit with a derived AC gate drive. Potentially, a transient signal on the gate could drive the circuit into a latched condition. The gate circuit therefore would need to ensure this condition is protected against using some suppression protection at the gate such as a diode or TVS. Alternatively, a transformer isolator drive circuit could be used, hence a latch in the DC circuit will decay to zero in the transformer and the AC signal passing through its zero crossing point will deactivate the SCR.

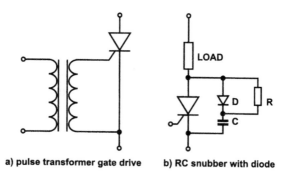

a) pulse transformer gate drive   b) RC snubber with diode

Figure 3.13

(a) Gate and (b) snubber circuits

3.4.2 TRIAC

The TRIAC is essentially a bi-directional SCR, hence AC signals pass during both cycles of the waveform. The driving of the TRIAC gate is essentially the same as the SCR and the above discussion should be applied.

3.4.3 Gas Surge Arresters

Gas (or plasma) surge arresters (sometimes referred to as gas discharge tubes, GDT) are constructed of two metal plate electrodes in a hermetically sealed package, usually ceramic, filled with an inert gas. When the potential across the electrodes is sufficient to ionise the gas a discharge occurs dissipating most of the energy creating the potential difference.

Gas surge arresters are used in similar applications to TVS devices. The main advantage of gas surge arresters over TVS or other protection components is the high currents they can handle (20 kA surge), hence large energy dissipating ability. Gas surge arresters are well suited to operation at constant high operating voltages and are most often applied to mains lines, three-phase power and telecommunication systems. The devices feature very low capacitance characteristics (<1 pF) and have typical operating times in the 1 µs region.

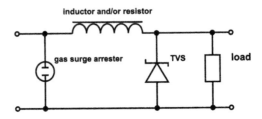

Figure 3.14

Full transient suppression using gas surge arrester and TVS diode

Compared with fast diodes for ESD protection the gas surge arrester can seem large and bulky and are usually only available in through hole packages. They do, however, offer some of the highest energy handling capacities of any protection device and are one of the few devices that can be used to protect mains and telecommunication lines against severe EMC threats such as lightning strike.

A problem on low voltage lines with the gas surge arrester is the high level of voltage at which the arrester can stop arcing (20–50 V); hence there may still be a large voltage present on the system even though the majority of the energy has been dissipated. When a system requires both large energy surge suppression on a low voltage line a combination of gas surge arrester and TVS diode or MOV would be used.

CHAPTER 4

INTEGRATED CIRCUITS

As the complexity of integrated circuits (ICs) increases, so does their ability to influence the electromagnetic compatibility (EMC) of any given application. Often there can be benefits as the closer proximity of components integrated into the silicon, shorter lead lengths (measured in microns rather than millimetres) and embedding of functions reduces the opportunity for radiated emissions and limits the power supply variations to within a few devices. There are downsides as well, as any problems that do occur are next to impossible to fix unless they present themselves for remedy on the device pins; also higher power requirements in a single device can produce greater current demand down a single printed circuit board (PCB) trace.

It could be argued that most of the electromagnetic compatibility (EMC) issues within ICs have to be dealt with by the IC circuit designers and IC suppliers; however, they may not be using or designing the device for the application you are attempting to produce, in fact you may be their only customer using the device in such an application where an EMC problem has occurred. Either way there are still a few basic EMC design issues that can be applied to even the most complex ICs to minimise the possibility of producing a poor EMC performance.

4.1 Bipolar or CMOS?

There may be a choice of semiconductor technology for certain IC applications, as combinational logic devices are available in various types of bipolar and CMOS processes. The majority of digital ICs are nowadays manufactured in CMOS with many analogue ICs following and except for specific application areas bipolar is certainly losing popularity. Mixed technology (BiCMOS) is gaining popularity in certain linear and mixed signal (digital and analogue) circuits and in ASIC designs. Unless specifically dictated by the circuit function, opting for a bipolar or unipolar technology will effect the EMC performance of a circuit.

The benefit is not always obvious, for example CMOS may not necessarily have the lowest power consumption as is often assumed. Bipolar can be run at very low-voltage levels, hence low power and does not have the switching power overhead of some synchronous CMOS devices. Static power consumption of CMOS may be lower, but at fast clock rates each CMOS IC introduces a large power demand on the supply. The power consumption of a clocked synchronous CMOS device is therefore dependent on the system clock frequency and device loading, at high frequencies the dynamic power demand of a CMOS IC may exceed an equivalent bipolar device. Consequently, at high switching speeds, it may be preferable to have bipolar circuits to reduce the conducted noise in a circuit and limit the amount of decoupling capacitance required.

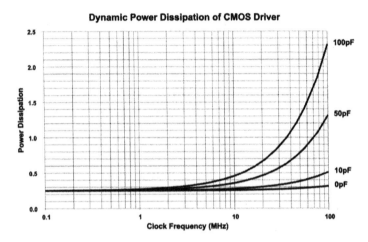

Figure 4.1

Change of CMOS dynamic power dissipation with capacitive load

The main downside to bipolar technology is its ability to demodulate radio-frequencies (RF). Both unipolar and bipolar devices can demodulate RF radiated fields due to the non-linear voltage–current relationships of the transistor structures. Unipolar has a square law dependency and bipolar an exponential law dependence and consequently produces a much better demodulator. In discrete circuits the demodulation is limited as even the smallest discrete device has a relatively large junction area, hence can demodulate only limited RF. Additionally, discrete circuit interconnect (PCB traces) often filters the signal from being conducted from a discrete device and being amplified within the rest of the circuit. In an integrated circuit the transistors have extremely small junction dimensions and interconnect between components is short, the demodulated signal can therefore be amplified within the IC. This is one method that RF can become injected into circuits, via a susceptible IC. The RF carrier itself appears as a DC offset on the susceptible signal, the demodulated signal could be anywhere in the frequency spectrum depending on the source. Common occurrences are audio frequencies from demodulated telecommunications and/or radio signals.

Examples of demodulated RF do exist, well known examples occur in professional audio circuitry (e.g. taxi radio broadcast pick-up on public address systems) and the effect is certainly more noticeable in audio analogue circuits. In digital systems the effect is less noticeable but could still upset system stability, the DC offset in particular can affect logic transition timing between susceptible and immune ICs. The effect of DC offset will be particularly problematic in low-voltage digital systems where even seemingly small DC offsets can produce signal jitter and instabilities, creating signal integrity and timing problems. The effect will be more problematic the faster the system clock rate and the faster the capability of the base logic used.

4.2 Integrated Circuit Packaging

ICs have almost as many package styles as they do functions. One common factor in most package styles, whether through hole or surface mount, is the assembly of the IC into its package. The IC is placed on to an island within the body of the package, the island is often attached to the lead frame in plastic-moulded packages or to the base material in ceramic packages. Bond wires are attached between each device pin and the IC bond pads. Although the leads from the package to the IC can be of differing length depending on their position at the package outline, the bond wires tend to be of very similar lengths, hence add a similar amount of inductance to each lead.

As with discrete packages the bond wire parasitic inductance is about 0.5 nH. Package inductance varies from pin to pin between 2 nH for shot central pins up to 10–20 nH for end pins in large DIL packs (e.g. 40-pin DIL types). Package capacitance varies by a similar amount, about 0.4 pF being a typical low value for short pins with about 1 pF for the longest types. Although there is some variation between plastic and ceramic material with regards to their dielectric properties, due to the dimensions and low values involved these differences appear insignificant in affecting the parasitic lead capacitance or inductance (Table 4.1).

Table 4.1 *Typical lead parasitics of IC packages*

Leads	Dual in-line (DIL)		Small outline (SOIC)		Plastic leaded chip carrier (PLCC)	
	L (nH)	C (pF)	L (nH)	C (pF)	L (nH)	C (pF)
8	6.3	0.68	3.1	0.35		
14	6.7	0.74	3.2	0.36		
16	6.9	0.77	3.4	0.38		
20	8.6	1.01	6.7	0.65	4.6	0.62
24	9.1	1.06	7.2	1.14		
28	9.6	1.14			5.8	0.72
40/44	11.0	1.25			5.9	0.77
64/68	17.6	1.51			6.1	0.80

As usual surface mount is the preferred style for EMC performance due to lower package parasitics. There is also reduced loop area with surface mount packages, particularly with quad flat packs (QFP). Some quad flat packs can offer 66% less loop area between trace pairs than comparable dual in-line package styles. Recent package styles, such as ball grid arrays (BGA) and tape automated bonding (TAB), can reduce package parasitics further due to reduced lead lengths and no bonding wires.

Even within a given package type there are EMC benefits to certain assignments of the pin connections. With DIL style packages for instance the pins closest to the centre are the shortest, hence these are the best pins to use for the supply lines (lowest lead inductance). Also packages with supply lines on adjacent pins make decoupling capacitor placement easier. The same arguments can be applied to other package styles, with the power supply rails and then the fastest signal traces requiring the shortest pins.

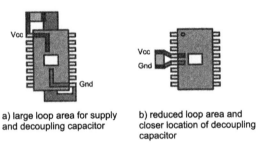

a) large loop area for supply and decoupling capacitor

b) reduced loop area and closer location of decoupling capacitor

Figure 4.2

Dual-in-line power supply loop area

4.3 Digital Devices

Many people believe that using a digital circuit, rather than an analogue solution, can circumvent some of their EMC problems as RF and EMC are considered analogue topics. Unfortunately, this view is very misguided, as one feature prevalent in digital circuits, sharp edges and square pulses, can generate a large amount of electromagnetic interference (EMI) with a large harmonic frequency content.

There is a lot that the designer can do with many digital circuits for minimising EMC problems, such as filtering and slew rate limiting, even when sharp edges and fast rise times may be inherently required for circuit functionality or are a feature of the devices in use. There are tricks and techniques that are useful to know and if a choice appears between a potential EMC threat and a potentially benign component, some background knowledge can help make the appropriate choice.

The main considerations are going to be based on the operating speed and rise or fall times of the digital circuits (it is mainly the rise times as most digital circuit transitions occur on the rising edge of the clock signal). The equation for use in estimating the potential harmonic threat of sharp pulse edges was introduced in the capacitor section. The equation gives the harmonic frequency content (f_{edge}) of a pulse with either a fast rise (τ_r) and/or fall (τ_f) time:

$$f_{edge} = \frac{1}{\pi\tau_r} \text{ or } f_{edge} = \frac{1}{\pi\tau_f}$$

4.1

This equation should be used to determine the frequency response required for decoupling capacitors (see capacitor section) and can be used to determine worse case tracking lengths when laying out a PCB. There could still be harmonic content present at up to 10 times this frequency if the signal is particularly square and overshoot and ringing could add additional harmonic signature.

The above equation will really only begin to predict the significant harmonic content from edges when rise and fall times exceed 10 ns. This generally occurs when clock frequencies exceed 10 MHz and by the time 50 MHz clocks are used the equation becomes essential to predict harmonic content. One problem of gauging the necessity of the equation is that many digital IC specifications only quote maximum rise or fall times. As far as EMC performance or problems are concerned the designer rarely knows the worse case conditions from the IC in use (i.e. the minimum rise/fall times). For example HC CMOS has a quoted maximum rise time of 10 ns, in reality 4 ns is more typical, which means that a logic capable of being clocked at a maximum of 32 MHz could introduce harmonic content up to 80 MHz into its signals.

Some digital designs can avoid many of the problems associated with these edge harmonics by having a level triggered design (i.e. trigger only when the clock is at a stable level rather than on a rising edge). As well as the high harmonic content of the transition, clock edges are also relatively noisy due to being connected to multiple devices switching all at the same time affecting the shape of the edge. By having a logic design that worked on stable levels, the effect of the edges on the circuit functionality is minimised and although the radiated and conducted emissions may be unaffected, the circuit itself is less susceptible. Level triggered logic is not as popular as edge triggered as it isn't quite as easy to implement, hence although a nice idea from an EMC standpoint, it can be impractical in many circuits due to the availability of level triggered ICs.

Owing to the threshold of the switching levels involved in digital circuits (typically 100 mV to a few volts) the components tend to have an inherent level of immunity to EMI that is not always possible with an analogue circuit. The high-voltage swing of both signals and clock edges, at relatively high rates does increase the potential emissions compared with a continuous analogue signal.

4.3.1 Clock Circuits

Potentially, clock circuits are the greatest EMC threat in a system. This may be somewhat unfair as it is not always the clock that causes the problem, but the circuits it tries to drive, including the PCB layout to these circuits. However, the most common noise problem from a digital circuit, either conducted or radiated, usually occurs at the same frequency as (or a discrete harmonic of) the clock.

There are some specific layout issues which are covered in the PCB section, but good grounding and adequate decoupling are two items that cannot be reiterated enough. Often other good EMC design practices are ruined due to simple errors in ground systems or inadequate decoupling. Additionally, a small bypass capacitor should be located with the clock driver circuit to reduce its load to the supply; a value as low as 100 nF should be adequate for most oscillators and resonator circuit supply lines.

It should not need stating that a designer should use the slowest clock speed possible, this is a general statement for any switching or logic device. If available use primary crystals or oscillators, derived or divided frequency parts will have the primary crystal frequency superimposed on the clock as well as the power line, which introduces unnecessary additional high frequency components of noise with much higher spectral power in the harmonics than would occur with a slower primary clock.

Slew rate limiting the clock is desirable, most logic ICs have a clock buffer on board and the slew rate-limited signal will be reconstructed as a square wave at the IC interface. Slew rate limiting only becomes a problem if using mixed technology logic ICs (e.g. CMOS and TTL) which have different switching thresholds. The threshold level itself is not the problem but the fact that the devices can have a slightly different switching point in time from the same clock edge. There are instances where the offset switching points can lower emissions, as the transitions cause a spread of power demand, similar to spread spectrum techniques (see below). Spread spectrum clock and spread spectrum device switching are, however, not the same and the latter can cause functional problems (signal integrity) and jitter on signal lines. Although slew rate limiting is a good idea in itself for reduced conducted emissions and reduced harmonic content it should be avoided with mixed logic types (although avoiding mixed logic families is preferable).

Limiting the available current drive to the known fan-out count also helps reduce noise problems, if available low output drive parts should be used. Overdriving the clock line wastes power and this power will inevitably find its way into noise or heat, causing a problem somewhere in the circuit. Distributing the clock to high impedance local buffers can also help as low circulating currents are distributed around the circuit with higher current demand localised to each sub-system or functional area. The clock line will need to be matched to the receiving line and terminating circuits may be required, but this should reduce reflections and ringing and hence overall EMI.

Dedicated clock driver circuits can also help in reducing the noise caused by clocks. The easiest example to consider is the multiple clock gate output circuit, this provides sufficient independent drivers and allows several independent lines to be driven. This arrangement permits each individual line impedance to be matched and reduces the possibility of cross-talk from one clocked line interfering with the other lines. A set of independent clock drivers with line-matched impedances also reduces

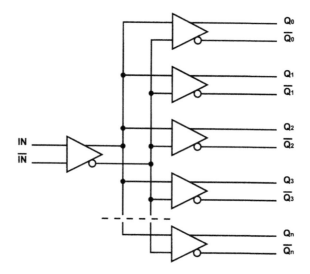

Figure 4.3

Clock distribution buffers

the possibility of clock skew at circuits furthest from the clock generator, and hence helps maintain functional synchronicity and signal integrity across the system.

A new approach to reducing emissions from the clock is the spread spectrum clock (SSC). This device gives a small spread in the absolute frequency of the clock by a few per cent centred around the fundamental clock frequency. Spreading the clock by a few per cent does not cause a problem with digital circuitry as the differences are only a few per cent and all devices see these small shifts at the same time, and hence remain synchronous. The overall system speed is unaffected as the time average is the same as the fundamental frequency. The advantage of this technique is that it reduces the amount of power at the absolute frequency by about 10 dB (again calculated as a time-averaged value), but incident power at any given moment in time remains at the same level as with a standard clock. The disadvantage is that there is a slightly wider range of frequencies being injected into the system, not only around the fundamental frequency but also around its harmonics. Feature sizes which will resonate at these small frequency changes can become problems that were not present with the original clock frequency and harmonics. The fundamental harmonic content average power is reduced, but edge rate harmonics are unaffected as the rise and fall times of the original clock are maintained.

One of the most useful types of driver or local buffer arrangement for reducing EMC on high-speed clock signals is the phase-locked loop (PLL) buffer. The PLL circuit can be used to synchronise two clocks if necessary, but more importantly from an EMC viewpoint the PLL allows a low speed, and even slow rise time, clock signal to synchronise a faster local clock. Using this arrangement allows a low speed signal,

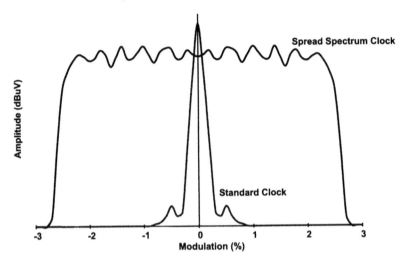

Figure 4.4

Spread spectrum clock emissions

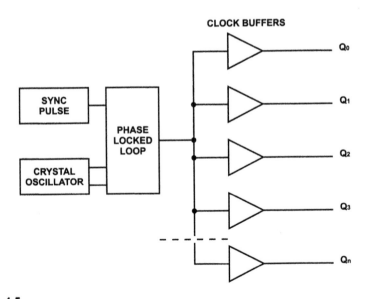

Figure 4.5

PLL synchronisation

say 1 MHz, to synchronise a series of circuits operating locally with a high speed clock, say 50 MHz. The advantages are relatively obvious, a low speed distributed synchronising pulse means that the longer tracks carry only the slower speed signals, whereas each circuit has localised short tracks carrying the faster busses and local high speed clocking. The major drawback is the additional cost of adding PLL devices to each circuit board or the complexity of design involved in adding a PLL circuit to a local programmable device or ASIC. There are many PLL circuits available that can synchronise high speed clocks on quite slow synchronising pulses and the idea should be borne in mind when using a multitude of processing circuits in the same system and requiring synchronisation between the circuits.

4.3.2 Combinational Logic Devices

This is probably the area of digital circuit design where a designer can most actively affect the EMC performance of the circuit. Combinational logic circuits are nowadays used mainly for what is termed as 'glue logic', that is to interface small sections of digital circuits together, either producing address decoders, bit counters or peripheral functions not directly available either in software or on a single IC, and where programmable logic is either too complex or too expensive. Usually, a relatively large amount of tracking and interconnect is required on the PCB considering the level of the function, and the designer has a range of logic families to choose their components from.

There is one base rule for choosing the appropriate logic family (as with all digital circuits), that is to use the slowest type suitable. This is a good rule of thumb and should be borne in mind during all component selection; however, there is a caveat in that the noise margin also significantly affects the EMC performance from a susceptibility standpoint. The lower the noise margin, the more susceptible a component is to external noise and triggering on EMI. Consequently, two families stand out as being especially good from an EMC viewpoint, CMOS and HC CMOS. There are additional design benefits from using these two types of logic, both are low power consumers, hence switching power demands on the supply are very light especially compared with TTL and LS types. It is also possible to run the CMOS and HC CMOS down to a 3.3 V supply rail, hence this type of logic gate can be used, unmodified, with the latest generations of low-supply voltage microprocessors.

There is of course a downside to all situations involving EMC where it appears you have a clear winner! The CMOS devices are more static sensitive from a handling viewpoint than the bipolar technologies (TTL, LS), but as most microprocessors are manufactured in CMOS processes this should not require additional precautions to a manufacturing line's existing electrostatic discharge (ESD) requirement. Of course the operation using CMOS is limited to clock rates of up to 6 MHz and up to 32 MHz with HC CMOS, so the range of applications can seem quite limited when processors with clock speeds exceeding 200 MHz are commonplace.

When using combinational logic in a circuit it should always be of the same family. Mixing logic families can cause excessive skew and jitter on clock lines and additional harmonics on signal and power lines. There is a potential for functional failure as well as increased harmonics when mixing combinational logic. From a susceptibility viewpoint the circuit is only as immune as its weakest component, using all HC CMOS with a single TTL gate still only provides the same level of immunity as the TTL gate can tolerate (a 0.4 V noise margin compared with 1.0 V for HC CMOS).

Table 4.2 *Comparison of common logic family parameters*

Logic family	Rise/fall time (ns)	Bandwidth (MHz)	Noise margin (V)	Decoupling capacitor (nF)
CMOS	100	6.3	1.0	0.47
LTTTL	20/10	20	0.4	0.33
TTL	10	32	0.4	2.2
HC CMOS	10	32	1.0	0.33
LS	10/6	40	0.3	3.3
ALS	4	100	0.4	2.2
S	3/2.5	120	0.3	1.5
F	1.75	180	0.3	1.5
ECL-10K	2	160	0.125	0.22
ECL-100K	1	320	0.125	0.22

Immunity can be improved in many combinational circuits simply by having all unused pins tied high or low, usually via a resistor. This is especially true of input lines, which can be left open but affect the functionality (e.g. ENABLE, PRESET and CLEAR lines). When logic lines are unterminated the inputs often float at approximately half the supply line, reducing the noise margin by half in many cases as the voltage excursion required to change state is halved. If a further DC offset is added, due to incident RF demodulation for example, the actual noise voltage required to trigger a pin could be only a few tens of millivolts. To minimise power consumption due to the termination the unconnected pins are usually tied to the ground rail via a high value resistor (4.7 kΩ for example).

4.3.3 Microprocessors

There are many types and designs of microprocessor available and not all the ideas given here are applicable to all variants. In particular there may be further techniques available to microcontroller types with on-chip memory and interface functions which are not covered in this section. When using microcontrollers or embedded microprocessors with other on-board functions, also check the section covering the additional on-chip functions discussed here as separate IC devices (e.g. memory, analogue-to-digital converters or ADCs).

Using the minimum clock speed processor is usually possible in applications where the device is being used as an intelligent controller, rather than in base computer designs (i.e. general purpose or personal computers, PCs). This is possibly a bit of an obvious choice as slower processors are usually also the lower cost types. The choice of processor speed may not always be easy to determine as some manufacturers may be selling faster versions, to shift off stock, under the slower version label as these faster devices meet the specification of the slower product. Another possible problem in trying to use a lower speed design is that many processors are being reduced in size by the manufacturer to achieve higher part yield per silicon wafer. Reduced feature sizes usually translate to faster transistors; consequently, although the processor clock rate may not increase, the rise and fall times may be increased and the harmonic content would therefore move up the frequency scale. Many of these changes may not be notified by the supplier, hence a microprocessor may initially be fine, then sometime during its production life cycle begin to suffer EMC problems. Although the part is still sold to the same specification, still labelled and packaged the same, internally the rise times have decreased and subsequently high frequency noise levels and susceptibility increase. If possible, and a higher speed processor is available, it is useful to try the faster device in the design with the lower clock rate to assess how faster rise/fall times will affect the circuit's overall EMC. If the PCB layout and other component guidelines are followed this will hopefully not be a problem.

Many applications may not have the option of using slower microprocessors, particularly where real-time operation is involved or where processing speed is an important factor. There are still many ways to reduce the potential for EMC problems.

Microprocessors with off-chip cache need to be sited very closely to the cache device, likewise with co-processors. The microprocessor and off-chip cache and co-processor communicate at very high rates, hence the PCB tracking length needs to be minimised to reduce radiated noise. Any clock circuit should also be sited close and all these ICs will need decoupling capacitors.

The electrical power required by microprocessors is increasing as their processing power increases and it is not uncommon to require a supply function (a regulator for example) very close to the microprocessor IC, with a separate bypass capacitor to reduce its effect on the DC supply to other circuits.

Microprocessors and other ICs which have multiple supply lines, either multiple voltage rails or multiple rails connected internally and/or on the PCB, should have a separate decoupling capacitor per power pin. The value can be quite low, say 4.7 nF per pin. If multiple ground rails are also available tie these together directly to the ground plane and/or place a ground image plane (covered metallisation) on the surface directly below the IC, this reduces ground impedance and radiated noise at the IC.

Where available a microprocessor which will generate the clock on-chip from an off-chip crystal or oscillator is usually preferred to a separate clock driver circuit. The processor is usually the highest power demand device at the clock frequency, hence locating the clock close to the processor ensures minimum drive demand at the clock

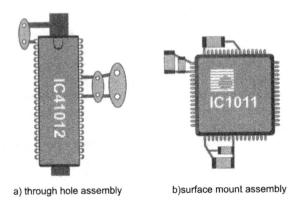

a) through hole assembly   b)surface mount assembly

Figure 4.6

Multiple decoupling capacitors

frequency distributed around the PCB. This also reduces the chance of noise corruption mis-timing the microprocessor. Noise on the clock will cause more serious problems to the processor than to any other circuit in the majority of designs.

As with most digital devices, pins which are input or mixed input/output are usually high impedance pins. These pins often either float to the mid-point of the supply rails or to an undefined voltage due to internal leakage paths. Either way high impedance pins are susceptible to noise and can register false levels if not properly terminated. Pins which are inputs and are not internally terminated, or terminated via externally connected ICs, need some high resistance (say 4.7 kΩ or 10 kΩ resistor) attached to each pin and ground to ensure a known state is present. This is particularly true of interrupt request (IRQ) and reset pins as a false condition on these pins due to noise will have a catastrophic effect on the circuit behaviour. A higher current consumption is often observed, particularly in CMOS devices, when input pins are unterminated as the input latch is half open, half closed resulting in a leakage current internal to the IC. Terminating high impedance input pins can therefore lead to a reduction in supply current, another benefit to both EMC and functionality.

Owing to the effects that interrupts have on microprocessor operation these are one of the most sensitive pins on the device. The IRQ could also be polled from devices at a distance to the microprocessor on the PCB, or even on a plug-in adapter or subsystem cards. Consequently, it is important to ensure that any line connecting to an interrupt request is protected against ESD transients. If possible a microprocessor should have level triggered interrupts as these are less sensitive to noise and, unless deliberately synchronised for a specific interrupt function, do not have to be synchronous with the clock. Bi-directional diodes, transorbs or metal oxide varistor terminations on the IRQ line are usually adequate for ESD and will help reduce overshoot and ringing without producing a significant line load. Again resistive termination will also maintain the IRQ line in a fixed state when not being polled.

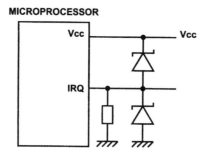

Figure 4.7

IRQ bidirectional diode protection

The microprocessor may be connected to a line driver for its bus interfacing. The bus is the second most likely source (after then the supply line) of an ESD transient reaching the microprocessor, and consequently if the micro is directly connected to the bus ensures that adequate ESD protection is provided at these connections. If the processor does not have adequate protection on-board then use ether a protected buffer as an intermediate interface or a bus protection diode array. The protection should be close to the bus lines rather than the processor itself. This type of protection makes commercial as well as EMC sense as usually the microprocessor is the most expensive component on a PCB, several hundred dollars of microprocessor can be protected for a few cents worth of diodes.

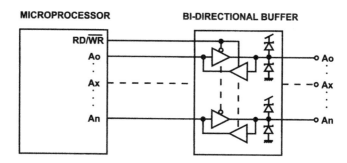

Figure 4.8

Bus buffer with ESD protection

4.3.4 Watchdog Circuits

A watchdog is a circuit which monitors the status of a digital line to determine if the system has crashed or gone into a closed loop. The circuit is usually used with microprocessor or microcontroller circuits where the processor is monitored for either a hardware or software crash (see Chapter 8).

The circuit works as a hardware timer, requiring a clock signal for timing and the monitored line to perform the counter reset before it times out. When the watchdog reaches a time out condition the device is made to poll the reset line of the microprocessor or digital circuit being monitored. Setting the watchdog time-out period is a difficult task and will depend on the system application, times between 10 ms and 2 s are common.

The watchdog circuit does not actually improve the immunity or susceptibility of a digital circuit, but can achieve a controlled recovery if the system is crashed due to an ESD transient or other EMI-related noise signal putting the system into an undefined mode of operation. The watchdog circuit is a safety feature for system recovery in the event of immunity problems, this is still useful in the case of excess EMI from environments the system is not intended to operate in (e.g. lightning strikes, short circuits on peripheral PCBs).

Microprocessors with on-chip watchdog timers are potentially the easiest to operate, all that is required is a few lines of additional code to handle the timing program and set the necessary counters. Some microprocessors can have the watchdog function programmed on-chip if a free running timer is available, and adequate reset programming is added to the code.

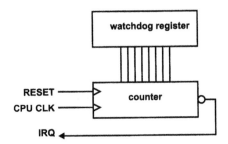

Figure 4.9

On-chip watchdog counter

Unless provided with an on-chip watchdog function, most circuits and microprocessors will require some programming and an output pin dedicating to the watchdog update. A weakness of external watchdog circuits is that they assume the error occurs within the microprocessor or digital circuit being monitored. An excess EMI problem causing false triggering on the clock could cause watchdog resets or false clocking (i.e. EMI problem with the watchdog itself). The watchdog could time-out faster than predicted whereas the digital circuit being monitored (the microprocessor) may in fact be immune to these noise problems due to other EMC protection measures.

Watchdog ICs are available as discrete functions with either programmable or fixed time-out periods. The circuit could also be produced using a simple divider circuit with a reset line. Watchdog circuits should be made as robust as possible so software programmable types are not recommended for ultimate robustness as their program could also be corrupted and produce false watchdog resets. The trigger for the watchdog toggle (clock) does not necessarily have to come from the system clock, in fact in most instances a slower toggle speed is preferable. The advantage of a slower toggle for the watchdog is that the toggle line could be filtered against high-speed transients with a simple low pass RC or LC filter without affecting the system clock.

Another feature which is advantageous to watchdog circuits is a fixed output reset switch (latch). If the reset line is tied to the main system clock rate, the duration of the reset in a noisy system may not be sufficient to activate the microprocessor reset line. A latched output reset arrangement will, however, require that the reset condition is acknowledged by the microprocessor and the watchdog is then also reset. If a feedback to reset the watchdog is not initiated then the microprocessor immediately sees a reset again after it has restarted.

An alternative to the feedback method is to have an astable output from the watchdog, this is easily and readily achieved when a standard divider circuit is used. In this circuit the astable output reset line from the watchdog is continually toggled high and then low until the microprocessor has restarted. The reset line will actually hold-off the start of the microprocessor for a fixed period (usually the time-out period) until the line is toggled and the processor can restart.

A level sensitive watchdog circuit has the advantage of a higher immunity to clock noise than edge-triggered devices; however, the disadvantage is if the watchdog is monitoring a signal line which has stuck in fixed state. On an edge-triggered watchdog reset this would not cause a problem as the timer would only be reset on the edge and would time-out as usual. With a level triggered watchdog a fixed reset level will prevent the watchdog from resetting the microprocessor, and hence the reset to the watchdog must be AC coupled to the monitored signal.

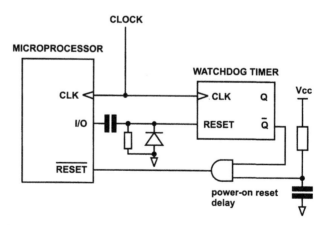

Figure 4.10

External AC coupled watchdog circuit

A good solution to the watchdog function is to use a dedicated output port of the microprocessor and add code to the system to set and clear this as the program progresses. This should then be coupled to the watchdog reset line via an AC coupling circuit. The advantage here is that the system itself requires two commands to reset the watchdog, both a set and clear, hence if a loop error occurs which includes only one of these commands the watchdog will still operate effectively.

The watchdog circuit is often forwarded by microprocessor suppliers as a way of designing for EMC, but it must be remembered that they do not improve the actual circuit EMC performance. Errors which are caused by software problems or hardware faults can also cause watchdog resets, so the circuit is not exclusively an EMI protection device. The benefit of the watchdog circuit is that the system should not get into a catastrophic mode of operation if an error does occur (EMI, hardware or software error), but will recover into a known reset condition (see Chapter 8 for further discussion). The watchdog circuit can also be combined with power-on reset delays, supply drop-out reset and other system reset functions with no additional program overhead as all these functions have a recovery cycle, which is in fact a program or system reset.

4.3.5 Programmable Devices

There are many programmable devices on the market today which are best addressed via other sections within this chapter. For example, one-time programmable (OTP) microcontrollers are covered by the notes on microprocessors. Programmable logic arrays (PLA) and field programmable gate arrays (FPGA) are probably best addressed via the combinational logic section.

One additional feature that some designers incorporate into programmable devices are test vectors brought to external pins used solely for test. Unless these pins are properly terminated they can act as unintentional radiators; however, connecting terminating resistors could significantly affect internal propagation delays. It would be recommended that designers only connect these vectors during evaluation and have the vector pin removed or internally unconnected in the production units.

Clock signals should only be admitted via a single pin unless there are two separate clock frequencies. Bringing in the same clock via different traces on the PCB could upset the timing and introduce clock beat frequency noise to the signals and supply lines.

4.3.6 Memory

Other than minimising the speed of memory (ROM or RAM) used to suit the circuit and siting the memory close to the processor to minimise PCB trace lengths there is little that can be done to improve the EMC performance of memory devices.

Read/write memory (RAM) is possibly one of the most susceptible ICs to the effect of transients corrupting data. Typically, data is stored as a charge on a transistor gate within the device, hence the possibility of a data, address or power line transient causing a corrupted data cell can be quite high. Dynamic RAM (DRAM) is slightly more susceptible than static (SRAM) as the cells are rewritten regularly helping to maintain the charge, but this allows potential corruption via the supply lines during the refresh multiple access to each cell. The difference in EMC susceptibility of DRAM and SRAM is very marginal and both will benefit from transient protection devices on data and power lines, especially if connected to bussed data and address lines without buffered access ICs.

4.4 Analogue Devices

Analogue circuits in general do not have the fast, square waveforms that a digital circuit inherits from its clock, consequently analogue circuits are often not as much of an EMI threat with regards to emissions as digital circuits can be. On the other hand analogue circuitry by definition is capable of reacting to small changes in signal levels and is therefore potentially more susceptible to EMI.

Keeping circuitry to within a specified bandwidth is usually easier with analogue circuits than digital, as rise times (or slew rates) are usually quoted as a maximum value rather than minimum, hence for EMC you should know the worse case condition for the component. If the worse case conditions for emissions are known they can be designed around. Similarly, the circuit itself will often have a maximum bandwidth of response defined by its functional description and adding simple low-pass passive filters operating outside this bandwidth will work without impairing functionality.

The threat from radiated EMI mainly occurs from the potential for demodulation of RF signals within the analogue circuitry and then amplification and/or transmission of a noise signal which, when demodulated, is within the normal operating frequency of the circuit. There are ways this can be limited by design and being aware of the way some components can respond to EMI.

Some analogue circuits are very susceptible to power supply variations, particularly precision amplifiers, audio circuits and ADC. Variations in supply rails can be superimposed on the signal, hence good decoupling and adequate supply line filtering is even more important for some analogue circuits than digital. Mixed analogue and digital circuits need good electrical and if possible physical separation to primarily prevent the digital circuit noise interfering with the analogue functions. Separate supply and ground rails would be recommended, to be connected only where either data is transferred or where the PSU supply enters the PCB.

4.4.1 Amplifiers

The amplifier is one of the basic building blocks of all analogue circuit functions, most other ICs discussed later will have an amplifier circuit of some type within their circuitry. The design of amplifier ICs can lead to the potential for a highly susceptible device, features which are good on the amplifier specification but bad for overall EMC performance include wide bandwidth, high slew rate, high input impedance and high gain. Amplifiers often feature a high sensitivity to power supply noise as the device amplification is ultimately limited by the power supply rails, also internal current sources can be affected by PSU noise, creating voltage offset errors.

Although losing popularity in many applications, dual supply rail amplifier ICs feature a higher common mode rejection ratio (CMRR) than single rail devices, providing a higher immunity to PSU noise and noise common to both amplifier inputs. Higher voltage supply devices tend to have a higher noise immunity for similar values of voltage gain and bandwidth.

The bandwidth of an amplifier is usually quoted at unity gain (i.e. as a 1:1 buffer rather than actually amplifying). When the bandwidth is reduced immediately gain is produced, hence within functional limits, increasing the gain of each amplifier circuit reduces is effective bandwidth. There is a counter argument to this however, as multiple stages of lower gain, but equal bandwidth, increases the gain fall-off characteristic by approximately 20 dB/decade per stage (e.g. a three-stage amplifier circuit can produce a roll-off of 60 dB/decade). A high rate of gain roll-off reduces harmonic content significantly without affecting the overall circuit signal gain. If a

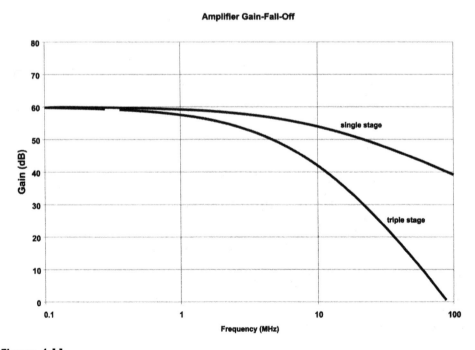

Figure 4.11

Gain roll-off of multiple stage amplifier

signal with a known high harmonic content (i.e. square or rectangular waveforms) is to be amplified multiple gain stages are better than a single amplifier.

The unity gain amplifier configuration can be used in applications where the source impedance of the signal, or signals, varies with frequency. A unity gain amplifier offers a very high impedance to the source, across a wide frequency range (wide bandwidth) and a low output impedance to the next stage of circuitry. Filtering a signal at a unity gain buffer amplifier can prevent noise from entering the target circuit at a very low cost. The signal response can still cover a wide band of frequencies without loss of signal.

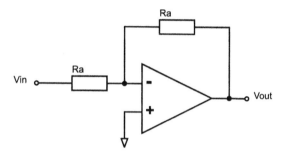

Figure 4.12

Unity gain amplifier

Guaranteed slew rate can be a useful feature, even if higher than required, as at least this allows a fixed filter to be used before or after the amplifier. Slew rate can easily be limited externally on an amplifier using a passive input filter or even in the feedback loop. Filtering in the feedback loop can be useful to change the gain-bandwidth product with frequency by changing the impedance of the feedback element (effectively increasing the roll-off within a single gain stage). If using feedback filtering care should be exercised on the possibility of resonance at higher frequencies causing a bypass of the amplifier itself, also any input noise still impacts the amplifier. Hence if filtering is to be applied it is always best applied to the input stage rather than the feedback circuit.

Current feedback amplifiers can provide a higher immunity than voltage feedback where these are available for the circuit function. Current feedback amplifiers feature a low input impedance, hence are less susceptible to input voltage noise than voltage feedback amplifiers. Another benefit is that the noise gain of the amplifier with current feedback is dependent solely on the feedback resistance (impedance), resulting in an improved noise to signal gain by choice of input resistance.

The main threat to radiated susceptibility in amplifiers, as in most ICs, is from demodulation of RF. In amplifier ICs the results can be more noticeable than in many other ICs as the demodulated signal can be amplified within the device after demodulation. This is a particular problem where dual supply rail amplifiers have a distinct advantage over single rail devices. In dual rail amplifiers the circuit forms a mirror between the rails and the RF signal should be demodulated in both sides of the circuit, hence appear as a common mode signal. Common mode rejection is high in dual rail devices and the demodulated residual signal is small and centred around the ground point, and therefore should appear as a very small noise signal compared with the input signal. This argument assumes the demodulated signal is within the bandwidth of the circuit function, where it is outside of the bandwidth filtering and control of the gain-bandwidth product should minimise out of band signal interference from any source.

Care needs to be exercised when using a standard amplifier which may be available from different manufacturers (e.g. the 741 operational amplifier). Different suppliers use different designs for the same function, consequently the EMC performance can vary between devices with the same functional specification. If possible and a range of suppliers are to be used, a sample from all the suppliers should be tried and tested against the EMC specification. The internal structure of an IC is going to affect its radiated EMC performance more than its conducted performance given equal CMRR, gain and other functional parametric performance, hence the need for testing in circuit.

4.4.2 Comparators

Comparators are generally insensitive to EMI as the function of a comparator is virtually digital. The output of the comparator is at the positive supply rail if the

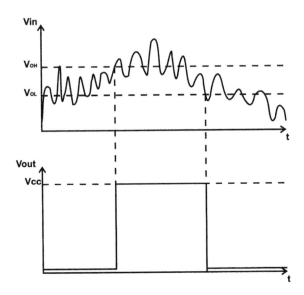

Figure 4.13

Schmitt triggered comparator

positive signal voltage exceeds the negative input voltage, otherwise the output swings to the negative rail. The output can reflect power supply noise, but this is unlikely to affect the device function significantly.

The main threat to circuit stability from EMI occurs at the threshold condition, when the inputs are close. If the input values are close and a noise signal is superimposed on one of the inputs the output can amplify this up to the power supply rail levels, producing a very high-power noise spectrum. A Schmitt-triggered device, which features a voltage hysteresis between the transition states improves immunity, this is directly analogous to the noise margin in digital logic ICs. Consequently, the greater the hysteresis voltage, the greater the immunity level.

4.4.3 Bandgap References

Bandgap references are inherently stable devices as they essentially consist of a forward conducting diode or p–n junction reference, some DC gain stage and biasing circuits (thermal compensation etc.). Consequently, after the supply has been applied the devices are stable and relatively noise free.

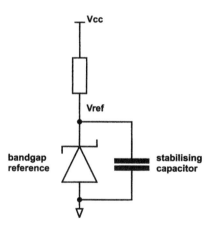

Figure 4.14

Stabilised bandgap reference

The main potential threat to the bandgap reference function is power supply transients. Some capacitance may be required at the device to provide additional decoupling and improve the device's response to transients' demands on the supply. It is unusual to require full ESD protection on these devices and decoupling of a few nF is usually adequate unless the device is in a particularly noisy environment.

4.4.4 Voltage Regulators

Voltage regulators are constructed from a bandgap reference, amplifier/comparator and power pass transistor in their simplest form. It would be expected that the device would therefore be relatively insensitive to EMC issues; however, situations do occur where the linear voltage regulator can create a high frequency internal oscillation. This is primarily due to high open loop gain of the comparator and pass transistor and potential for positive feedback at RF frequencies due to phase shift.

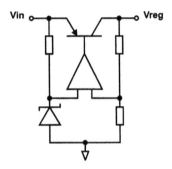

Figure 4.15

Basic voltage regulator circuit

The oscillation problem usually occurs with inadequate decoupling close to the regulator pins. The problem is particularly noticeable if the load is some distance from the regulator, hence there is an inductive lead between the regulated output and load. The output voltage from the regulator leads the load demand, and unless stabilised with quite large capacitance (10 µF/A) the internal comparator oscillates on and off at relatively high frequencies (20 MHz oscillations have been observed on a 5 V regulator supplying only 200 mA in a TO92 package).

Not all regulators suffer this as much as others; however, there is no obvious way to determine this from the individual device specification. Generally, higher current devices are slower to react and consequently do not suffer such high oscillations, having sufficient internal junction capacitances to stabilise the circuit. Devices using Schmitt trigger comparators are also less prone to a high frequency oscillation.

Programmable regulators (i.e. those with a separate voltage feedback pin) are usually easier to stabilise and require smaller capacitors. The stabilisation is added to the feedback/programming pin, and therefore usually a 10 nF ceramic capacitor can be adequate as the current load for the feedback pin is small. It is also possible to sense the voltage at the load and feed this back to the regulator, hence the lead inductance problem is moved from a high current lead to a low current lead reducing the noise problem.

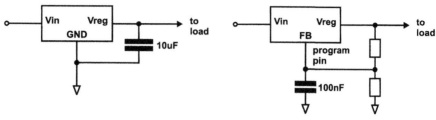

a) stabilising capacitor when load is at a distance from the regulator

b) stabilising capacitor at feedback on programmable voltage regulator

Figure 4.16
Stabilised linear regulator

The most recent developments in linear regulators are for lower voltage drop versions of the standard three terminal devices, termed 'low drop-out' regulators (LDO). These are generally no better or worse from an EMC standpoint to their higher drop-out parents. However, operating at a lower input-to-output differential can reduce the feedback loop instabilities, hence potential oscillation may be easier to prevent with a smaller capacitor.

As with other IC commodity components, regulators are common devices available from many suppliers. Some suppliers exhibit the oscillation problem quite readily whereas others have never experienced the problem.

4.4.5 Serial Interface Devices

Although most often used for communication between digital systems, interface standards such as RS232 and EIA-485 are considered analogue circuits. Unlike many analogue devices they are relatively insensitive to power supply noise due to the digital nature of their signalling schemes. However, as interfaces to external systems they are potentially a source for noise entering the system, and therefore may need higher level of protection than is apparent from their parametric performance specification.

There are many standards for serial interfacing but these can be classified into two groups: single ended and differential. In single-ended systems the signal return is via the circuit ground or reference rail, in differential signalling a separate return signal, or complementary signal line is employed separately from the circuit ground reference. Consequently, with regards to EMC, the differential signalling scheme has advantages over the single-ended scheme; however, differential schemes are more expensive to implement, primarily due to the additional cabling requirements. Table 4.3 lists four of the most popular serial interface standards and their characteristics.

Table 4.3 Serial interface standards

Parameter		RS-232-C	RS-423-A	RS-422-A	EIA-485	Units
Mode of operation		single ended	single ended	differential	differential	
Number of drivers and receivers		1 driver 1 receiver	1 driver 10 receivers	1 driver 0 receivers	32 drivers 32 receivers	
Maximum cable length		15	1200	1200	1200	metres
Maximum data rate		20k	100k	10M	10M	bits/s
Maximum common mode voltage		±25	±6.0	−0.25 to 6.0	−7.0 to 12.0	V
Driver signal level	Minimum	±5.0	±3.6	±2.0	±1.5	V
	Maximum	±15.0	±6.0	±5.0	±5.0	V
Load resistance (capacitance)		3.0k to 7.0k (2500)	450	100	54	Ω (pF)
Driver slew rate		30	*	**	**	V/µs
Driver resistance		300 (power on)	—	—	—	Ω
High impedance state		300Ω (power off)	±100 ($V_o = ±6.0$ V)	±100 (-0.25 V ≤ V_{cm} ≤ 7 V)	±100 (-7 V ≤ V_{cm} ≤ 12 V)	µA
Minimum receiver resistance		3.0	4.0	4.0	12.0	kΩ
Receiver sensitivity		±3.0	±0.2	±0.2	±0.2	V

*Determined by cable length and data rate.

**Determined by IC used.

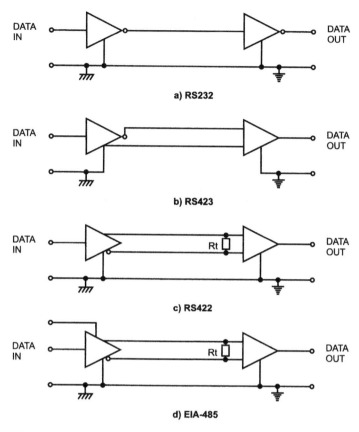

Figure 4.17

Interface standards wiring schemes

Single-ended schemes are similar in principle to most digital logic signalling schemes, except that higher voltage levels are used and signalling is usually between a positive and negative voltage (e.g. with typical RS232 circuits signal levels are between +12 V for logic '1' and −12 V for logic '0'). Single-ended serial interfaces have a much higher immunity to interference than logic circuits due to these higher signal voltage levels; however, as the cables are in external uncontrolled environments they can be expected to see higher levels of noise than circuits boards within a system. Single-ended schemes are rarely terminated resistively, hence sit in a high impedance state and rely on the high voltage level of the signals as a barrier against false triggering. The problem of high impedance on single-ended lines is that interference can produce a voltage offset on the ground line, even the signalling itself can create ground bounce offset, hence reducing the noise immunity of the signal line. The ground levels of two interconnected systems can also be offset by a significant DC level due to independent reference points or even feedback ground current, which again reduces the signal switching level in absolute terms and therefore reduces the noise immunity.

Differential schemes offer high levels of immunity with lower signal levels as these operate on the difference between two signal levels and not an absolute voltage level. Having differential signals also increases the level of common mode rejection from the signalling system as interference on the cable superimposes on both signal carrying lines and sums to zero at the receiver circuit. The most popular differential schemes (RS-422 and EIA-485) also require resistive termination of the data line for correct operation. As well as resistive termination reducing reflection and ringing, resistive termination also reduces the potential for cross-talk and interference as the lines are linked by a low-impedance connection and can sink noise voltage or current induced in the cables.

Emissions from serial interfaces can be reduced by the use of slew rate limited signalling devices. These devices operate using a controlled rate of switching, producing slower edges on signals and, consequently, lower harmonic content in their spectra. For example, an EIA-485 interface operating at 250 kHz signal rate using an interface IC without slew rate limiting can exhibit significant noise spectra above 10 MHz without significant fall-off of power density. Using a slew rate limited device for the same signal and the harmonics are almost extinct by 5 MHz (Figure 4.18).

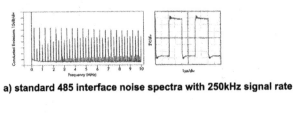

a) standard 485 interface noise spectra with 250kHz signal rate

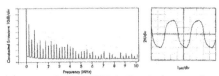

b) slew rate limited 485 interface noise spectra with 250kHz signal rate

Figure 4.18

Noise spectra of standard and slew rate limited EIA-485 interfaces

Slew rate limited ICs are available to drive most interface standards and RS232 has a slew rate limit in its specification (see Table 4.3). Slew rate limited ICs are often restricted on the maximum transmission frequency at which the interface can transmit data (slew rate limiting only affects transmission rates). In the majority of cases this should not be a problem as the rates are still high, but in speed critical interfaces slew rate limited switching may not be suitable.

Filtering can be used for both immunity and emissions reduction. Basic passive filters are adequate to produce some slew rate limiting on the transmission and filter noise at the receiver. A common mode choke arrangement is typically used on differential signal lines with a simple differential filter for single-ended schemes (see section on inductors for calculation of component values).

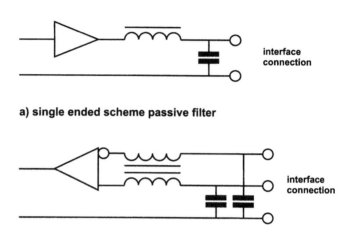

a) single ended scheme passive filter

b) differential scheme passive filter

Figure 4.19

Filtering serial interface lines

Another technique commonly used to improve the EMC performance of an interface circuit is to isolate the interface. There are several off-the-shelf isolated interface hybrid devices available which in many cases provide single chip solutions to the interface isolation circuit. Alternatively, the interface can be isolated using discrete components, using opto-isolators for the data and either a DC–DC converter or transformer and switching power supply device for the supply line. Having an isolated interface significantly increases the noise immunity as even ground-induced noise becomes common mode at the interface, also ground offsets between systems are reduced as the interfaces are no longer referenced to the system ground.

As an accessible port in a system, ESD is also a significant potential threat to an interface circuit. It is also a likely access point of ESD to other in-board circuits, hence the interface is the best location to provide ESD protection. Single-ended systems are usually easiest to protect using a single bi-directional transient voltage suppressor device. With differential signalling several devices may be required on either line and the loading capacitance will have to be minimised for the higher speed signalling lines. Some of the best methods are to include the ESD protection components within the connector (see section on connectors). Many IC manufacturers recognise the threat of ESD on interface circuits and integrate protection into the interface IC itself; however, the ESD pulse could still create interference on other in-board circuits and discrete protection at the connector terminals provides a degree of protection for the entire system, not only the interface IC.

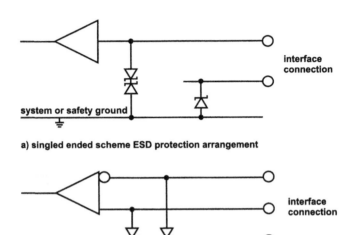

a) singled ended scheme ESD protection arrangement

b) differential signalling scheme ESD protection arrangement

Figure 4.20
ESD protection devices at interface connection

4.4.6 Data Converters

Analogue-to-digital (ADC) and digital-to-analogue (DAC) converters are often an interface between an enclosed digital system (e.g. microcontroller) and an external analogue process (e.g. measurement of some physical parameter such as temperature). Consequently, as with serial interfaces, there is a potential for both susceptibility and emissions via the converter circuits.

There are additional reasons for considering the noise content of the analogue signal due to potential false data values being generated by the converter. The ground rail and supply lines for the analogue and digital sides are also usually separated and only connected either at the IC pins or even on-chip in some instances.

The level of noise that can create a problem is the analogue voltage resolution (V_{AR}) and is relatively easy to calculate for ADC/DACs. It is related to the digital resolution (i.e. number of bits, n) and the full scale analogue voltage swing (V_{FS}).

$$V_{AR} = \frac{V_{FS}}{2_n}$$
4.2

Therefore, for an eight-bit device with a 0–5 V full-scale range, the analogue resolution is 19.5 mV and noise above this level could produce false data. A 16-bit device with the same analogue full-scale range is sensitive to only 76 µV of noise. Hence higher bit converters will be more sensitive to noise than lower bit devices, or alternatively the analogue range could be increased to compensate for each additional bit.

The speed of ADC/DAC converters is usually much lower than the digital processing circuit and filtering of the analogue signal is easily implemented with passive components. Alternatively, some analogue signal processing (e.g. sample and hold, amplification) can be applied directly to the analogue signal and include the filtering in the active circuit.

The signal can even be digitally filtered, for example the digital value can be repeatedly read from the ADC (or written to the DAC) and the numerical average taken, or more complex error correction algorithms can be applied. The digital filtering techniques can only be applied when the analogue sampling rate is lower than the digital sampling rate (i.e. the analogue signal rate of change, or slew rate, is slower than the digital sampling frequency). Digital sampling of this type can be achieved with additional software rather than electrical components, so the filtering cost can be written off in the development stage and, providing the additional code fits in available memory, there is no component count increase.

One common feature to almost all types of mixed signal converter that significantly affects the EMC is the stability of the voltage reference device used within the circuit. The reference is frequently on-chip, but available to an external pin for decoupling. A relatively low value of capacitance is required, typically 10–47 nF being adequate, but if available the manufacturers' guidelines should be followed. The reference is used to determine the analogue signal value by comparison via other on-chip circuitry (parallel resistor ladder, successive approximation or dual slope) and often has a low value; therefore, with a high-bit converter the level on analogue noise on the reference required to cause false data is very small (tens of microvolts typically). If an external voltage reference is used the notes above for stabilisation should be followed and the device should be sited close to the converter and over the same analogue ground plane.

The type of converter used is usually dictated by application so the EMC performance is secondary; however, as with most devices the slower the device operation the lower the potential for EMI problems. With signal converters the fastest types are usually parallel or flash converters, these present a relatively large instantaneous load on the supply on the conversion clock edge. Flash converters should only be used where their speed is essential (e.g. video digitising circuits and digital storage oscilloscopes or DSOs).

Successive approximation converters are used for medium speed conversion, typically up to a few million samples per second (MS/s) in processing and instrumentation applications. This type of converter offers a reasonably even load demand on the supply as the on-chip conversion progresses in single steps. Consequently, the noise generation on the supply line is relatively low. Dual slope types are typically the slowest and the lowest supply noise generators.

The synchronisation of signal converters and control logic is usually not necessary for correct operation. The controller reading the converter is often operated at a much higher rate than the analogue conversion process so that data can be read or written, processed and the result communicated to other devices while the next conversion is being performed. Even high-speed video DACs and DSO ADCs operate on signals in the digital domain at a much higher rate than the analogue signal is sampled. This allows the converter to be deliberately operated either asynchronously or delayed from the digital system clock.

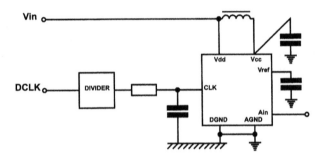

Figure 4.21

RC time delayed clocking

Full asynchronous conversion has little benefits as regards EMC and the asynchronous conversion cycle can create beat frequencies with the digital circuit. However, an offset clock signal has a benefit to the converter as the conversion edges are slightly offset from the digital clock edges, hence the noise from the digital circuit supply occurs at a different instant to the analogue sample. With adequate decoupling on the converter this slight clock offset can significantly reduce the effect of the digital supply noise affecting the analogue signal. The delay can be generated simply using an RC network which will also slew rate limit the conversion clock signal into the converter, again benefiting the EMC performance.

Circuit layout can prove critical to full functional performance of the converter. Similar principles to interface ground separation need to be applied to the analogue and digital supply lines (see Chapter 6). The supply for the analogue and digital circuits should be separated at the device. If the analogue supply is derived from the digital rail it should be filtered using an inductor or ferrite bead and have a separate decoupling capacitor on the analogue supply input pin. The analogue ground should be separated and be used to provide a guard ring and/or surface ground fill for the analogue circuit sections up to the analogue signal connector.

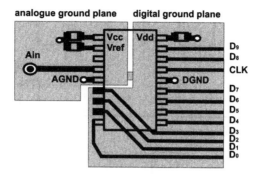

analogue ground plane   digital ground plane

Figure 4.22
PCB layout for ADC/DAC

ESD protection used for serial interfaces can also be applied to a signal converter analogue input/output connection.

4.4.7 Power Supply Components

A modern trend in power supply design, and hence in associated ICs, is to use a switching technique to produce step-up and step-down supply rails at a high efficiency. The linear regulator (see above) is still popular, but these newer switching regulator circuits are becoming more commonplace as an alternative high-efficiency replacement.

The switching regulator is a significantly higher EMC threat than a linear regulator as it has an inherent oscillation, hence noise source, for its functional operation. Coupled to the controller/oscillator IC is usually a magnetic component, either a transformer or and inductor, possibly an external power transistor and several other passive devices to complete the circuit. There are a multitude of potential EMI generators within a switching power supply circuit and whereas most of the other analogue ICs are potential EMC victims, the switching power circuit is a potential EMC aggressor.

There are several choices that can be exercised over the IC as well as the peripheral circuitry. Although a high maximum switching frequency can equate to smaller magnetic and filter capacitors, maintaining a primary switching frequency below 150 kHz can reduce conducted noise significantly within the regulatory band and make filtering easier. There are also choices of switching topology that will affect the noise characteristics of the completed circuit.

The most basic controllers feature a simple fixed frequency switching circuit to drive a magnetic component and the output is determined by the value of the magnetic component (either the value of an inductor or the turns ratio of a transformer). This type of supply is an unregulated circuit and noise content is easy to filter and often low due to the low-power levels these circuits can handle.

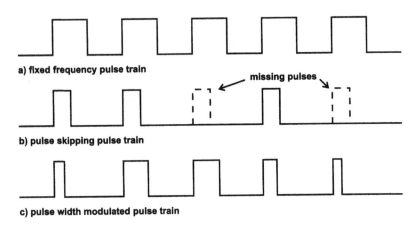

Figure 4.23

Various switching technique typical pulse trains

Pulse skipping controllers work by having a fixed pulse width and frequency and omitting occasional pulses to regulate the output voltage from the circuit. This topology is relatively simple and having a fixed frequency and pulse width gives a great advantage to the noise characteristic, the maximum repetition rate is known and can be filtered and most other noise is a sub-harmonic of this, hence a low EMC threat.

A popular technique for higher-power levels (20–100 W) is pulse width modulation (PWM). The controllers for this switching technique regulate the output voltage by controlling the width of the pulse, usually with a fixed repetition rate (frequency). Another factor which can be controlled or limited is the duty cycle, from 0% to 100% of the repetition rate. Limiting the minimum duty cycle to a known value helps to minimise the high frequency content, as very narrow pulses create high-frequency harmonics. PWM techniques are more EMC aggressive than pulse skipping or fixed frequency techniques due to the uncertain nature of the pulse train and the difficulty in filtering over a wide modulated pulse range.

Soft switching techniques are another method of limiting noise content of switching circuits. Soft switching is not necessarily simply slew rate limiting the pulses, but usually entails allowing the magnetic circuit to saturate and the energy to decay to zero before the next switching cycle is initiated. These types of circuits are also known as resonant mode as they make use of the self-resonance of the magnetic circuits to determine pulse timing. The techniques offer low losses as the switching circuit itself does not have to remove any energy from the magnetic element for the next switching cycle. Figures determined from simple unregulated 50 W switching converters suggest soft switching techniques can reduce conducted noise by as much as 30 dB.

Circuits switching large current loads and power levels exceeding a few hundred watts may use zero crossing controller ICs to achieve low conducted noise. These controllers have a topology which has a state during the cycle in which either no current flows (zero current crossing, ZCS) or no voltage differential is applied to the magnetic element (zero voltage crossing, ZVC). This gives a low conducted noise as the transition of switching occurs during a zero energy state on the magnetic element (the technique is also popular with motor controllers). Zero crossing techniques can be used with other control modes and are particularly useful for high-voltage circuits.

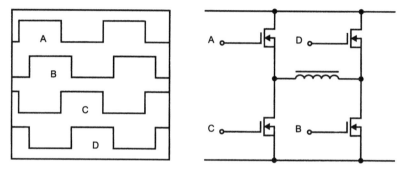

Figure 4.24

Zero crossing H-bridge transformer driver

The main threat for radiated noise in power supply circuits occurs from fast switching edges and the magnetic circuit element (see section on inductors and transformers). Even low frequency switching ICs can produce fast rise times for the drive and these may need limiting externally. PCB layout and grounding schemes will need considerations and ground return loops within the PSU circuitry should be treated as high speed lines and tracked close to the supply feeds for minimum loop area and maximum coupling.

As with linear regulators, the potential for internal resonances can occur in the feedback loops in power supply circuits if the open loop gain of the feedback system is very high. The feedback loop may need its own decoupling and capacitive compensation to reduce its operating speed. The requirement for fast response to transient load demand in modern supply circuits makes the limiting of the feedback circuit response difficult to justify in many instances, particularly for microprocessor circuits. Using simple feedback circuits which have a near digital response (on/off) may be better for fast load demands than a high gain analogue feedback system (e.g. slope compensation).

A recent approach to active noise suppression in power supply circuits is the power supply damping circuit (PSDC). This is essentially an active frequency-dependent shunt resistor which is high impedance at low frequency and low impedance at high

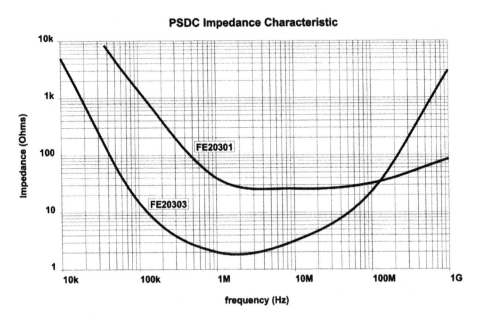

Figure 4.25

PSDC impedance analysis

frequency. Hence noise on the power supply is shunted to ground between certain frequency limits. The device does not offer as low an impedance as some capacitors, but the frequency range over which the low impedance extends would require multiple parallel capacitors to achieve. The range of devices is relatively restricted, primarily aimed at 5 V DC supply rails. The cost is also significantly higher than the cost of a capacitor, but the device does offer a higher level of noise suppression for certain power supply applications.

CHAPTER 5

ELECTROMECHANICAL AND HYBRID COMPONENTS

There are components used within any electrical circuit that do not easily fall into the categories so far covered. Some of these may even be considered as not strictly being an electrical component (e.g. wire); however, as they are often constituent parts of a circuit and may influence the electromagnetic compatibility (EMC) they deserve to have some coverage.

There is also a large range of electromechanical and mechanical devices which are commonly used in electrical and electronic systems, from basic plugs and sockets to relays. These can influence the system level EMC to a great extent, especially mechanical switches where contacting mechanisms can cause both conducted transients and radiated electromagnetic interference (EMI) if not controlled by design.

5.1 Passive Electromechanical Components

5.1.1 Wire and Cabling

Despite most designers best attempts to remove wire from a system by placing all tracking on a printed circuit board (PCB), there are very few systems which do not feature some wiring. Often switches, indicators, PCB-to-PCB connections and power supply leads are common even with high-density PCBs. Where possible it is

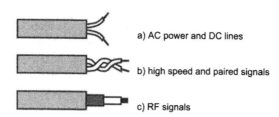

a) AC power and DC lines

b) high speed and paired signals

c) RF signals

Figure 5.1

Cable types by application

advisable to minimise wiring within a system, with a PCB track the location and routeing of every signal and supply rail can be controlled by the designer and once fixed will not vary once the circuit goes into production. With a wiring arrangement there are possible differences in lead length and in placement within the system. A relatively benign power supply unit (PSU) lead could be routed accidentally past a microprocessor clock and end up with a high-frequency noise linked directly into the PSU and, therefore, every circuit in the system.

There are some basic principles that can be applied to all wiring schemes, minimising the loop area being one of the easiest and most effective. This not only involves ensuring that the leads are as short as possible within the system, but also that where there is a signal in a lead, its return lead is close. Signal and return leads should be paired within a cable. Minimising the lead length minimises lead inductance and maintaining close coupling with the return signal maximises capacitive coupling between signal and return, which can also minimise coupling to other circuits. If possible wiring should be segregated to known paths so that potential interference is minimised due to the positioning of the wires. Using a conduit or series of cable ties within an enclosed system helps maintain a known and controllable level of EMC performance from the cabling system.

Twisted pair wiring (sometimes referred to as unshielded twisted pair, UTP) is probably the best low cost solution for signal wires and PSU connections. In twisting the wires the loop area is reduced close to zero, within the preciseness of the number of twists per unit length. The capacitive and inductive coupling between signal and return is also maximised hence common mode noise should self-cancel. Unshielded cables of this type are commonly used in data communications systems up to 4 MHz with little degradation in signal integrity or EMC problems. The frequency range can be extended up to a few tens of MHz by shielding the twisted pairs (STP), again data cabling systems operating at 16 MHz have been successfully installed with this type of wiring system without problems.

In shielded cabling schemes the shield is usually connected to a single ground within the system. Where grounds are isolated connecting at several places is not usually a problem, but with non-isolated ground connections, a ground loop could be induced between circuits or systems that can offset one of the ground potentials. An offset ground potential can make a system more susceptible to radiated EMI as the noise margin at one end of the system is usually reduced and conducted noise between systems or circuits is greater. If the coupling mechanism is known or suspected, then a grounding scheme based on the coupling can be used. If capacitive coupling is injecting interference then a single ground at one end should be used (usually the signal source), if magnetic coupling is introducing noise then grounding the signal shield at both ends is recommended.

Ribbon cable, or other unshielded multiple signal cabling, should include several return or ground lines. The inclusion of additional ground lines maintains closer coupling between signal and return paths. A ratio of three signals to one ground (3:1) should be attempted to maintain low noise by ensuring a relatively good coupling between the signals and ground, the best approach would be alternating signal and ground lines;

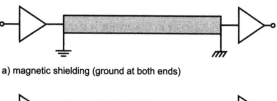

a) magnetic shielding (ground at both ends)

b) capacitive shielding (single ground at source)

Figure 5.2

Shielding connection

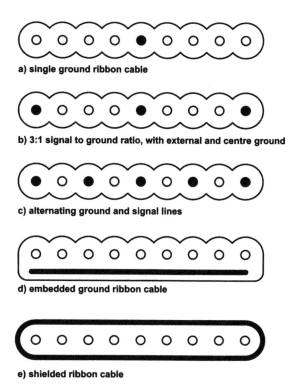

a) single ground ribbon cable

b) 3:1 signal to ground ratio, with external and centre ground

c) alternating ground and signal lines

d) embedded ground ribbon cable

e) shielded ribbon cable

Figure 5.3

Ribbon cable types

however, this scheme is rarely feasible in high-density systems. Having a ground on either end of a ribbon cable or in the centre is a minimum design constraint that should be pursued to maintain a constant coupling level. One benefit of ribbon cable over loose wiring types is that the location of the signal and ground is known to be fixed relative to each other, hence some control is exhibited over the cable characteristic impedance, even when additional ground schemes cannot be implemented.

Shielded ribbon cable would be a preferred medium if a large number of signals is to be interconnected (such as co-processor daughter boards and bussed memory and data lines). A lower-cost alternative is the embedded ground ribbon cable, this is analogous to the microstrip method of PCB tracking and provides a controlled impedance cable suitable for most high speed digital signal lines (data and addresses busses with clock frequencies greater than 20 MHz).

When very high speed signals are to be transmitted (above 100 MHz), either internally in a system or to the 'outside' world, high quality shielded cables will eventually become mandatory. The most common form is coaxial cabling which is usually specified by its characteristic impedance (50Ω or 75Ω being typical values). This type of cabling has been successfully used in short runs (under 50 cm) up to 18 GHz, although a more practical limit for most systems would be 500 MHz. The outer cable shield is usually a ground return path and by maintaining the centre conductor at a constant fixed distance gives a highly controlled impedance characteristic over a wide frequency range. As the frequency increases the effectiveness of the outer conductor as a shield is reduced and at frequencies above 300 MHz care should be exercised over cable placement, length and routeing. The use of the shield for the return signal in coaxial cabling systems has a limited effectiveness and STP cabling is often equally effective up to several tens of MHz at a much lower cost.

If ultrahigh speed (above 1 GHz) and high integrity shielded cable is required, tri-axial cable with two outer shields is available. The outer most shield is system ground, the second braided shield is the signal return or local ground and the high-speed signal is conducted along the inner wire. This type of cabling is expensive and should be used only where absolutely required. The connection of each layer of shield can become a problem in itself with three concentric signal connections. There are high quality plug connectors (N-type) for this type of interconnection system and signal integrity is usually excellent up to microwave frequencies.

With cables that enter and exit a system maximum care with the interface is imperative. In situations of cabling that leave the system, the potential for noise injection and emission is much higher than in enclosed cabling systems. Where possible shielded cable should be used for signals that exit the system, with attention paid to the termination of the shielding at the point of exit and entry between systems.

Pig-tail shield terminations should be avoided wherever possible, and when unavoidable should be made a minimum length and terminated close to the cable inlet. Pig-tails create an effective ground inductance and hence can act as a high impedance on the shield to high frequency signals. Complete 360 shielding contact should be made within the plug housing or at the cable terminating clamp.

a) pig-tail termination

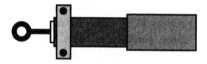

b) 360 degree cable clamp

Figure 5.4

Avoid pigtail terminations

A feature of wire which can potentially be used to the advantage of the EMC of a system is the effect of skin depth (δ). This is the effective depth from the wire surface a signal will penetrate due to its frequency (f_s). The result is a higher impedance to high frequencies in single strand cables (sometimes referred to as 'Bell core') and the wire itself can act as a low pass filter. The magnitude of skin depth in metres is given by the equation:

$$\delta = \frac{1}{\sqrt{\pi\mu\sigma f_s}}$$

5.1

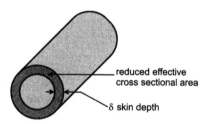

reduced effective cross sectional area

δ skin depth

Figure 5.5

Effective skin depth in wire

Where σ is the conductivity of the wire (58.8E6 Mho/m for copper) and μ is the permittivity ($\mu_r = 1$ for copper, $\mu_o = 4\pi E{-}7$ H/m). Hence for copper wire the skin depth expression can be written in terms of mm as:

$$\delta = \frac{66.1}{\sqrt{f_s}}$$

5.2

Therefore a signal of 100 Hz will encounter skin depth impedance effects in wire gauges greater than 13.2 mm (twice the skin depth). A 1 MHz signal will experience increased impedance in wire gauges greater than 132 μm and at 1 GHz the signal will only penetrate to a depth of 2.1 μm from the wire surface.

It is effects such as the skin depth which limit the ability of cabling to carry high frequency signals. The limit for conducted emissions in the European EMC directive is 30 MHz, at frequencies above 30 MHz the skin depth in copper is less than 30 μm. The most common copper PCB tracks have a thickness of 35 μm (1 oz copper) and typical multicore cable consists of wire with a diameter greater than 80 μm; consequently, this frequency would appear to be a reasonable limit to the basic unimpeded signal carrying capability of most conductors used in circuit construction.

Skin Depth in Copper

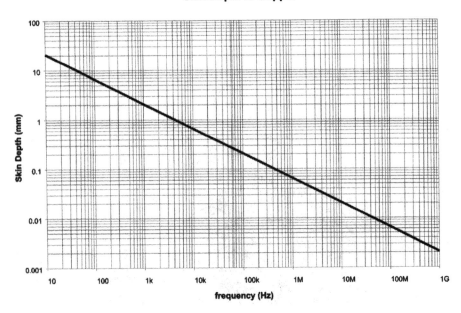

Figure 5.6

Skin depth in copper

5.1.2 Connectors, Plugs and Sockets

Connectors, plugs and sockets have long been recognised as a potential weak link in a system's EMC chain, as the way signals and power link in and out of a system are usually the main sources of EMI entering and exiting the system. Coupled with cabling methods, the methods of connecting system signals at the PCB are varied, from internal data channel (IDC) connectors on ribbon cable to fully shielded UHF

N-connectors on tri-axial cable. Obviously shielded connectors and plug housings are preferable; however, they may be overkill in many situations and a lower cost unshielded connection may be adequate, especially for internal system interconnections.

Hardwiring two systems would be the preferred choice for the exchange of signals, as once the desired EMC level is achieved, the cabling will not be able to create problems at a later date. Unfortunately, this is rarely possible and a system designer has to try and specify the types of cabling, length of interconnect and other parameters to ensure the system integrity and EMC performance is not compromised once the system is installed and in use. There are not always clear guidelines on connecting systems other than to use 'quality' cables.

Some general principles can be applied to all types of connectors, both for high speed and low speed signal connections. The blades, beams and tails within a connector should be maintained as short as possible, this helps to minimise inductance at the connector (approximately 1 nH/mm per pin). Try and use sockets which have a complete encapsulant coverage of the connector beams throughout the connector, this maintains an even impedance throughout the connector beam. Connectors which feature a plastic casing only around the socket outlet or at the PCB contacts have an irregular impedance profile to high speed signals, this type of construction is more prevalent in right-angled connector assemblies.

At low signal speeds often the connector characteristics can be ignored; however, as signal frequencies exceed 10 MHz and rise times reduce below 2 ns the short electrical lengths in a connector can no longer be ignored. The characteristic pin impedance of a connector (Z_c) is governed by its characteristic inductance (L_c) and capacitance (C_c) given by:

$$Z_c = \sqrt{\frac{L_c}{C_c}} \qquad\qquad 5.3$$

The inductance is minimised by having short blade and beam lengths in the connector. The capacitance is minimised by increasing the space between pins, typically capacitance is the dominant parameter in connectors $(C_c \gg L_c)$. The characteristic socket impedance can therefore be selected to match that of the PCB tracks or driver/receiver circuits, or alternately the PCB tracking can be controlled to match the connector. Matching connector, cable and PCB track characteristic impedances provides a method of transmitting high speed signals with a minimum of reflection, skew and signal loss. The tracking to a high density connector may need to feature serpentine tracking on the closer pin connections to match propagation delays into the connector.

If the cable itself is unshielded there is little point in using shielded plugs; however, grounding the socket could help with the EMC of the driver/receiver circuits on the PCB, particularly with respect to electrostatic discharge (ESD). A metal shielded connector can exhibit up to 6 dB of noise rejection at 100 MHz and 40 dB at 1 GHz with suitable grounding of the shield even with exposed, but recessed, pin connections.

Even unshielded connectors and cabling can have some simple filtering applied at the socket using some of the ferrite absorber inserts which are appearing on the market. These absorbers are a gasket of ferrite and act in the same way as a ferrite clamp (see section on inductors). They are designed to fit specific socket forms and may not be available for all sockets on the market. The absorber can be retrofitted if a suitable type is available, hence can help solve high frequency signal problems on systems which are at the end of their design cycle.

Another socket add-in which works in a similar way to the ferrite absorber is the Chipguard™ ESD suppressor. These are a metallic strip that fits into the socket form or behind the socket on the PCB connection. The device operates in the manner of a transient voltage suppressor (TVS) between each pin and the socket housing. Consequently, these devices require a grounded socket housing or local ground connection to operate effectively. Again these are available as a retrofit solution for systems where ESD protection has not been adequately designed in. (Chipguard™ is a trade mark of Bourns Inc.)

ESD in particular is a serious potential problem for socket housings that are user accessible. Most peripheral interfacing and inter-system connections are user accessible as it is the user who is going to connect the devices/systems together, hence these require ESD protection at the socket. Using a grounded socket with a metal sheath may be all that is required if the socket aperture is small, the discharge will naturally go via the path of least resistance and a bare metal grounded sheath offers very low resistance. With wider aperture sockets additional ground connections throughout the socket help with ESD as well as noise, but ultimately sensitive lines will require correct ESD design at the component level (e.g. metal oxide varsitor, TVS or Chipguard™ products).

All shielded cables will require shielded plugs and shielded socket receptacles. There is not much point in shielding your cable without correct plug and socket shielding as the benefits to the signals in the cable will be lost. The cable shield should be terminated through 360° in the plug housing to mate with the plug shielding. The interconnection of cable shield and plug shield is relatively straightforward with high frequency connectors such as Bayonet Neill-Concelman (BNC) and N-types;

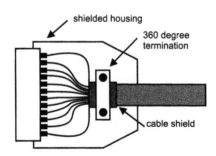

Figure 5.7

Terminate shielded cable in shielded plug housing through 360°

however, with D-type plugs and other arrangements a great deal of care needs to be taken with the shielding terminations to maintain shield effectiveness.

Any plug housing for interconnecting shielded cables should have a 360° connection to the cable shield. Pairing back the insulation and clamping with a collar is usually the best and most convenient method unless there is a specific fit for the plug (e.g. BNC connectors). As with hardwired connections, pigtails of the shield connection within a plug housing, even a shielded plug housing, should be avoided due to the additional inductance this adds to the shield, reducing its effectiveness at high frequency.

Back plane sockets are usually unshielded and often only accessed during system assembly and during upgrades or system extensions, and consequently do not require the ESD protection that user accessible connectors require. High density sockets are used and these will often be carrying several high speed signals (data, addressing and clock synchronisation). The signal arrangement within the pin structure of the connector can be used to improve its EMC. In particular, adding more ground connections within the connector not only reduces the inter-board ground impedance, but lowers the characteristic impedance of the connector pins and cross-talk within the socket. Optimal cross-talk reduction is achieved from having a ground adjacent to a signal on four sides; however, this arrangement requires a very large number of pins available for ground contacts. A more conventional arrangement is to attempt a 2:1 signal to ground connection arrangement within the connector. When using differential signalling schemes the connector should carry signal pairs in adjacent pins, with staggered ground connections.

Sockets are one of the few components that are still predominantly through hole and not surface mount. The main reason for retaining through hole assembled sockets is

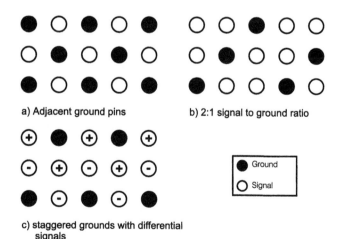

a) Adjacent ground pins   b) 2:1 signal to ground ratio

c) staggered grounds with differential signals

● Ground
○ Signal

Figure 5.8

Optimal grounding arrangements for high density connectors

the mechanical strength this gives to the socket on a PCB, especially if the socket is frequently accessed (i.e. has many plug-in, pull-out operations during its lifetime). Other reasons are the difficulty in placing sockets with automatic machinery, hence loss of one of the main benefits of surface mount device assembly, and as the system is going to enter a cable the need for minimised inductance on the socket is often obviated. There are surface mount sockets on the market and for automatic production these may be advantageous, especially in internal interconnections where the cabling will only be accessed once at system assembly. There are no real EMC benefits to using surface mount sockets, in fact through hole can offer some advantage as a ground plane around the connection pins can be applied on the PCB, which will slightly improve the decoupling of the signals.

5.1.3 Fuses

Fuses as components are not inherently EMC sensitive, unless the incident energy to the fuse is sufficient to overload the fuse wire. In most cases it will be the placement and mounting of the fuse that could cause an EMC problem, rather than the component itself.

As with power switches, the fuse must not be positioned so that it effectively circumvents any installed shielding or power line filtering. If the fuse is to be user accessible it is best placed near the power inlet and have short connecting wiring to the inlet socket. Filtering should therefore be between the fuse and the power supply in the system.

In-board fuses are best left as non-user accessible, as bringing the fuse to the enclosure surface can create apertures by which EMI can radiate out of, or into, the system.

Fuses are effectively single wires in space, hence could act as aerials for EMI in the fuses locality. Fortunately, the fuse wire is usually relatively short and only susceptible therefore to very high frequencies. Being a single-stranded wire its skin depth is quite poor at these frequencies and hence it is not a very good receiving aerial. Nevertheless, the fuse positioning should be away from high frequency sources.

5.1.4 Switches

Mechanical switches have two mechanisms for creating EMI, although determining which is present in a given system is difficult, but fortunately both can be suppressed by the same circuits. When the contactors of a switch come close to contacting the potential across the gap can create an arc for a very short period of time and this causes a transient current pulse. When the contactors make initial contact the spring mechanism can cause the switch to 'bounce' and produce an intermittent contact for a very short time period. Both the arcing and switch bounce create a fast rise time current pulse into the circuit connected to the switch, which can create conducted and radiated noise, especially as switches are often connected to a circuit via a wiring loom rather than directly attached to a circuit board, hence have unintentional radiating aerials attached.

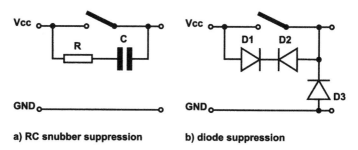

a) RC snubber suppression   b) diode suppression

Figure 5.9

Switch contactor suppression

The usual method for reducing contactor noise on a switch is to place an RC snubber network across the contacts. A transient protection diode could be used, providing the switch noise energy can be handled by the diode. The transient noise generated is across the terminals of the switch and can be bi-directional, hence it is usually better and less expensive to dissipate the noise in an RC snubber network than a diode. As with any transient protection it should be located at the transient source, the snubber should be directly across switches' external contacts and not on a PCB at the end of a lead to/from the switch.

The choice of value for the resistor can be calculated if the peak current is known (I_{pk}) and given the supply voltage (V_{cc}):

$$R \geq \frac{V_{cc}}{I_{pk}}$$ 5.4

The peak current is not always known and 10Ω is a common value used. Likewise the required damping capacitance (C) can be calculated to retard the rate of voltage rise across the switch terminals. A rise in excess of 1 V/µs is required in a DC circuit to initiate arcing, knowing the load current (I_L):

$$C \leq I_L \frac{dV}{dt} = I_L \; (\mu F)$$ 5.5

This calculation can result in a relatively large capacitor for large current switches, however, the capacitor also has to be able to handle the rapid rise time of the arc/bounce (typically 8–20 ns). Although this equation gives an upper boundary condition, capacitance values chosen from a capacitor's impedance curve are also valid and values of 1 nF in a ceramic dielectric are typical for this application.

Positioning and mounting options for switches can have a major influence on their EMC performance. Mounting a switch in a panel often requires leads to/from the switch, hence increases the opportunity for receiving radiated noise, and radiating contactor noise. Where possible direct PCB mounting with panel connection would be preferred.

There are situations where the EMC design of a PSU has been completely compromised because the mains switch has been connected to/from a front panel and hence internal radiated noise can circumvent the PSU shielding by coupling to the switch leads. Likewise PSU noise circumvents the in-line filters and can radiate from the leads into the target system. Wherever possible mains switches for a PSU should be located close to the PSU, and if shielding is included, within the PSU shielding itself.

5.1.5 Component Sockets

In general component sockets should be avoided. They add a further 1–4 pF of additional parasitic capacitance per pin and a few nH of additional parasitic inductance. Certainly, clock and oscillator circuits or their peripheral components, should not be socketed.

Sockets are relatively common with programmable devices and microprocessors in upgradeable systems. It would undoubtedly be better for EMC not to use a socket, but it can be a feature (i.e. upgradability). Low contact resistance, high quality contacts and usually tight forced fit type sockets are best.

Zero insertion force sockets (ZIF) usually have the longest leads both from the socket to the PCB and within the jaws of the ZIF mechanism. ZIF sockets can add over 10 nH of series inductance and 8 pF of inter-pin capacitance per device pin and are really only suitable for test purposes, not where high speed system integrity is required.

Where sockets cannot be avoided, component sockets featuring the shortest lead lengths should be used and a few more nF of local decoupling capacitance added to compensate for the additional socket inductance. Many surface mount sockets are now available and these would be the preferred choice where socketing is unavoidable.

5.1.6 Heatsinks

Although not an electrical component or necessarily connected directly to the circuit, heatsinks can act as both electromagnetic (EM) shields or capacitive coupling for RF signals.

Where heatsinks are to be used with active components such as diodes and transistors which are electrically isolated but thermally coupled to the same heatsink, the heatsink provides capacitive coupling between components attached to it. If any of the components are subject to high-frequency signals, or are particularly sensitive to EMI (e.g. small junction area devices) there should probably be an additional screen between the heatsink and the device to reduce coupling (similar to the Faraday shield used to reduce transformer winding coupling). A thin sheet of copper with a local component ground connection, between the heatsink and the active devices, should reduce capacitive coupling between components common to the heatsink by 6 dB or more.

Alternatively, it may be possible to electrically connect all heatsinks to ground. The impact on leakage current due to capacitive coupling will need to be considered, especially if the ground used is the mains earth connection. This arrangement can increase the shielding of the complete circuit without explicit shielding being used, hence at no additional cost.

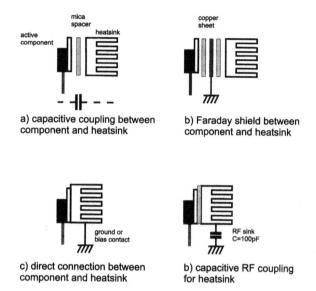

Figure 5.10

Heatsink connection options.

Wherever possible metallised areas, such as heatsinks, should not be left electrically floating and should be connected to either safety ground, circuit ground or static supply rail. Areas of floating metallisation are potential RF coupling mechanisms and can create resonances and radiate internally to the system as well as transmitting internal RF out of the system. Where connecting directly to an internal circuit connection is not possible, connection via a small capacitor (e.g. 100 pF) to an internal ground gives a sink for RF current incident on the metallisation. The heatsink may be connected to a suitable rail within the circuit, for example if the component case is a live circuit connection, such as the collector on a transistor, then the heatsink can be connected to this. If using a live circuit connection the heatsink should be connected to a static bias or other non-signalling part of the circuit to reduce its effect as a radiator/aerial and to prevent it inadvertently loading a signal line with additional capacitance. If connecting the heatsink to a supply rail its location in the system and any isolation or safety aspects should of course not be compromised.

5.2 Active Electromechanical Components

Active electromechanical devices will by the nature of their operation be EMC aggressive. The function of creating a mechanical motion from an electrical signal usually entails creation of a magnetic field to move a permanent magnet or magnetise a metal actuator. There is therefore a deliberate generation of an EM field, as well as the additional potential EM disturbance of physically moving magnetic and metallic objects. Fortunately, the EMC aggressive nature of these devices has long been considered in the design of control circuits and much is already known on minimising the EM disturbance created by these devices.

5.2.1 Solenoid

The solenoid is possibly the simplest of active electromechanical devices, essentially consisting of two components, a coil and an armature. It is the connection of the armature to a mechanical assembly that gives the solenoid its function (e.g. electromechanical valve, electrical door lock).

The main EMC threat posed by a solenoid is the energising and de-energising effects of the coil winding. Often there are many turns in the coil and quite a large current is required to activate the solenoid, hence there is a relatively large supply surge required when the coil is energised and a large back electromotive force (EMF) when the current is switched off. The energising current can be compensated for by the use of sufficient bypass capacitance in the control circuit. The solenoid is not usually operated above a few kHz so relatively large capacitors can be used successfully to provide both bypass and decoupling.

The de-energising back EMF is the main threat to conducted and radiated EMI, which can have a rapid rise time and produce a large voltage transient spike. A catch diode across the terminals is a minimum protection that should be incorporated and an RC snubber can also be used, which will help reduce overshoot and ringing in the coil. The larger the coil assembly the larger the diode will have to be to handle the energy from the back EMF, but often larger coils feature slower rise times, hence can utilise slower diodes. It would be expected that the cost of an appropriate protection diode would be insignificant compared with the cost of the solenoid itself.

Any protection circuit should be mounted at the solenoid terminals. The solenoid can be remote from the control circuit and connecting cable could radiate any transients, hence, as usual the transient needs suppressing at its source.

Solenoids are not very susceptible components due to the large amount of energy required to activate them, and hence do not usually need protecting from transient noise themselves.

The magnetic fields generated by solenoids tend not to be as problematic as in their electrical component equivalent (the rod inductor) as their application usually places them remote from other components that could be sensitive to the magnetic field. The field is still present and placement of other circuits, including the control circuit of the solenoid, close to the coils and armature should be avoided. Circuit placement does not have to be too remote (a few cm) to reduce the influence of the magnetic field significantly. If the control circuit is located on the solenoid shielding may be required to minimise magnetic field coupling to the components and hence back into the system.

5.2.2 Electromechanical Relay

The electromechanical relay is a solenoid with the armature connected to a switch. As such it has two potential EMC problem areas, the inductive nature of the coil and bounce of the switch contacts. Relay coils need to have some transient and/or overshoot protection from the back EMF generated when the coil is de-energised, as with the solenoid above. Usually, an RC snubber network or catch diode is used to prevent the coil transients interfering with other circuits in the system. These transient protection circuits need to be located at the relay coil terminals to reduce possible radiation of the transient by connecting cable or tracking.

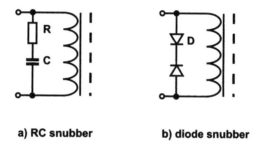

a) RC snubber   b) diode snubber

Figure 5.11

Solenoid or relay coil suppression circuits

Switch contact bounce requires the same suppression as applied to standard mechanical switches already covered in this chapter.

A modern alternative, and better from an EMC standpoint, is the solid state relay (SSR). The SSR has the advantage of a low or non-inductive drive element and a non-mechanical switch action (see hybrid components).

5.2.3 Electric Motors

The nature of electric motors, a rotating EM machine, indicates that these will be very EMC aggressive. Much of the mechanical housing design for motors is intended to reduce radiated EMI, hence some EMC considerations have already been installed. There is still more that may be necessary to ensure the motor and its control systems are EMC friendly.

There are many types of motor available but two classifications can be applied: brush type and brushless. Motors that feature brush connection to the motor windings, via a rotating commutator on the motor armature, usually generate much more electrical noise than a similarly specified brushless type. This is due to radiated EMI from the sparks caused by the constant contacting action of the motor coils and brushes via the commutator. Although diode and capacitor suppression at the brushes can help reduce the conducted noise, other than maintaining the brushes and commutator in good condition little can be done electrically to reduce the radiated noise. Good screening around the commutator head and brushes will be required with a good connection to a ground or mains earth (depending on the application of the motor).

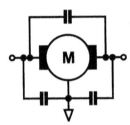

Figure 5.12

Motor commutator suppression capacitors

Brushless types do not feature an electromechanical contacting commutator, and hence have a much reduced radiated EMI. The electrical switches creating the commutation can be relatively soft in their switching action (typically field effect transistor or insulated gate bipolar transistor switches are used), hence commutation noise can be maintained to relatively low levels and control can be exercised over the commutation. Adding diode or RC snubbers to each coil is also a relatively easy task and can further reduce the EMI. Each coil on a motor can be treated with regards to EMC in a similar way to the coil in an electromechanical relay.

The switching topology applied to control the motor can also influence the EMC of the system. Simple single-sided drive circuits are potentially the worse for EMC but are often the lowest cost control topology. In low-power systems (e.g. servo or small DC motors) these can be operated reasonably well with simple diode or snubber damping. As the power level required by the motor increases so does its potential to create EMI. Some of the best topologies for minimising conducted noise at high power levels are the zero crossing techniques, either zero current (ZCS) or zero voltage (ZVS) switching topologies. These techniques deliberately reduce either the voltage across the windings, or the current in the coil to zero prior to commencing the next energising phase. The circuitry can be relatively complex involving up to four transistors per motor phase, hence can be expensive to implement and control.

The zero crossing circuits do not produce any direct supply-to-ground connection during the switch cycle. Consequently, the conducted noise and transients generated by this topology are related solely to the energy requirement of the target circuit (motor coils in this instance). The suppression circuitry for this control topology is therefore relatively straightforward and covered by the usual bypass and decoupling rules (see section on capacitors). Owing to the lower level of conducted noise, the

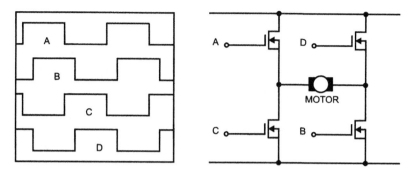

Figure 5.13

H-bridge drive for DC motor

radiated circuit noise is also lower and in a motor circuit the suppression of the coils themselves is the dominant EMI problem when using a zero crossing control circuit.

5.3 Hybrid Components

Hybrid components can include virtually any circuit function, some can even be described as sub-assemblies. If designing a hybrid component the sections so far covering individual components and the PCB sections should be consulted. Here the few commercially available parts which are sold directly as components and therefore considered as commodity items are covered, but whose construction dictates that these are really hybridised circuits.

5.3.1 Opto-isolators

Although strictly a combination of two discrete devices, opto-isolators are sold as a single functional device. The part usually consists of a light-emitting diode (LED) and a light-sensitive bipolar transistor in a single package. Internally the devices are physically separated by an optically translucent filler between the devices. Considerations applied to signal diodes and small signal bipolar transistors can be applied to the opto-isolator. As bipolar devices are used, the part is relatively insensitive to ESD.

The opto-isolator can be used as a good noise barrier for digital signal isolation as the bipolar output can be operated in a switch mode so only the digital content is passed and any superimposed noise on the signal is not transmitted (including power supply noise). This is particularly useful at interfaces between systems, the ground loop is broken and any noise picked up by the cabling appears as a common mode noise at the cable terminals, hence can be relatively easily cancelled either solely by

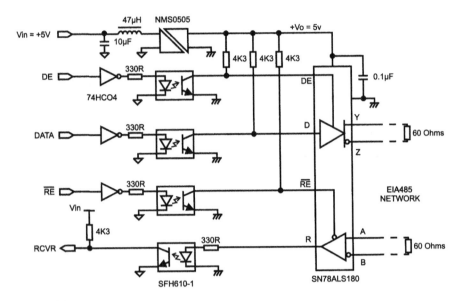

Figure 5.14

Opto-Isolated EIA-485 data interface

the opto-isolator or using additional common mode chokes. The use of a safety ground for ESD and transient protection devices is made easier as the signal ground is isolated from the system ground. The down-side to this is cost, opto-isolators are expensive devices compared with other discrete components and a local isolated power supply (e.g. DC–DC converter) will be required to power the signal side.

There are a range of linear opto-isolators now available, still containing diodes and transistors as the transmission devices. This type of component is useful for analogue signal isolation and the breaking of analogue ground loops. The main drawback compared with the digital opto-isolator is that noise can be transmitted across the barrier, particularly power supply noise superimposed on the signal. If using a linear opto-isolator adequate decoupling and filtering of the supply feeding the input signal stage is essential.

5.3.2 Solid State Relays

The SSR is gaining popularity as a direct replacement for many electromechanical relays, certainly in low-current switching applications (<10 A). There are several benefits over the electromechanical device with respect to the EMC performance, the fact that the switching action is 'soft', that is electronic not mechanical, and the input is not necessarily inductive (although this is not always the case) are beneficial. Additional benefits of the SSR over the electromechanical device is that it is mechanically quieter as a result of having no moving parts, hence has no audible switching noise.

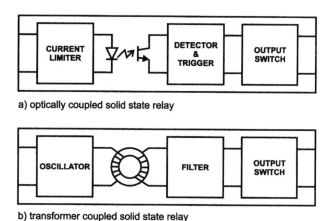

a) optically coupled solid state relay

b) transformer coupled solid state relay

Figure 5.15

Solid state relay

The construction of the SSR usually consists of an opto-isolator or transformer isolator on the control side, and a transistor acting as the switch. The current drive requirements for both the transformer and opto-isolator types are significantly lower than an electromechanical relay, there is consequently a reduced conducted noise potential from the SSR. The SSR is slightly more susceptible than the electromechanical relay, but is still much less susceptible than most ICs and discrete components, hence susceptibility is not a great problem unless operating close to high voltage machinery.

If using a transformer isolated SSR transient protection for the transformer coils back EMF and overshoot may be required. As the drive requirements are lower, the generated transients are also lower, but also slightly faster. Consequently, a small (fast) low-cost silicon diode can be used to reduce the transient, placed at the SSR control terminals. This suppression circuitry may be included within the SSR assembly.

The switch action of an SSR does not contain any mechanical bounce, hence it is possible to avoid applying any protection on the switch terminals. If some transformer bounce from the control side is affecting the switch characteristics the control circuit should be either slew rate limited or have a snubber circuit applied. As the control side typically handles much lower power levels than the switch side it is easier and more cost-effective to apply any ESD/transient measures at the control pins.

5.3.3 DC–DC Converters

These board level components offer a neat method of isolating the DC power supplies of numerous circuits. The DC–DC converter consists of a silicon switching circuit coupled input to output via a transformer, usually ferrite cored, and an output rectifier and filter arrangement. Hence the circuit has a switching element and this is

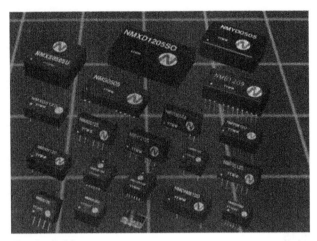

Figure 5.16

Board level DC–DC converters, courtesy of Newport Components Ltd

the main EMC concern of this type of circuit as the switching action is a potential noise generator.

As a circuit isolator the device can offer many EMC benefits for conducted noise suppression. At a relatively low cost (compared with PSU redesign) a system can have an isolated DC supply between all circuit functions, hence conducted interference via the DC supply rails is significantly reduced. The devices are frequently used to isolate control circuitry for high-voltage circuits from their digital signal lines, similarly at interface functions which use opto-isolators for the signal isolation DC–DC converters are used to power the interface and in isolated analogue-to-digital interfaces where low noise analogue sensing circuits are in use. Hence DC–DC converters are already used in many circuits where conducted noise problems have traditionally been encountered.

There are many different topologies for providing the switching function and these significantly affect the conducted noise generated by the DC–DC converter itself. As a general rule, the simpler the device the easier the generated noise is to deal with. Fixed frequency converters are easy to filter as they have a characteristic frequency, given in the data sheet, which changes very little with loading or environmental conditions (i.e. temperature). Fixed frequency converters tend to have no sub-harmonics to the switching frequency and are thus easy to filter with a simple LC or RC low pass filter circuit. Although the filter should be designed around the primary switching frequency, the switching and/or rectification scheme may produce higher power spectra in the second and subsequent even harmonics, which results in a higher attenuation of the noise from the filter because the power spectrum is shifted one harmonic above the primary frequency. Fixed frequency converters are primarily used in the lower power ranges (sub-5 W) and for fixed input voltage circuits.

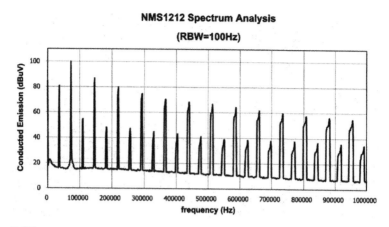

Figure 5.17

Conducted noise spectra of fixed frequency DC–DC converter

In circuits where the output load varies considerably during circuit operation and converter efficiency needs to be maximised, and in circuits where the input voltage can vary significantly due to poorly regulated input supply, different switching techniques are used. Pulse skipping techniques miss certain pulses in the switching cycle to regulate either the input or output voltage; consequently, these contain sub-harmonics due to the differences between the time the pulse is applied to the transformer from cycle to cycle as the output loading or input line (supply) voltage change. There are also a wide range of upper harmonics with pulse width modulation (PWM) techniques, where the pulse width is varied to again regulate either the input or output voltage. Both of these techniques offer a greater EMC threat than the fixed frequency fixed input converter as more complex filtering is required and sub-harmonic spectra are possible.

PWM is a greater EMC threat than pulse skipping as this potentially could contain very high frequency content from narrow pulse trains. This usually occurs with low output loading and/or with a high input voltage. The frequency content is not always specified and often any noise data will be quoted at full output load, hence not under the worse case condition for this type of converter. The worse case condition occurs with minimum output load at maximum input voltage as this situation should create the narrowest pulses. Measurements on this type of converter also need to be made using a quasi-peak detector circuit as the average noise level may be low due to the low cycle rate, but the EMC threat of very narrow pulses is greater than a high repetition rate of wide pulses.

Pulse skipping techniques often feature a fixed pulse width and maximum repetition rate. Consequently, its highest frequency is often relatively easy to identify and most sporadic noise spectra will occur in the lower frequency ranges. Low frequencies are not as much of an EMC threat as high frequency, hence despite the loss of

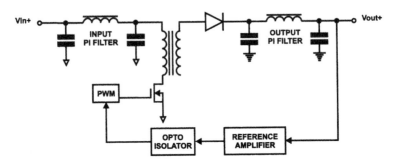

Figure 5.18

DC–DC converter PWM topology

predictability of the frequency spectra from a pulse skipping converter it is often as easy as a fixed frequency converter to filter with respect to EMC.

There are some switching techniques which will assist in reduced noise on conducted lines. Soft switching methods usually have a limited slew rate switching action and hence a reduced harmonic noise content. In low power converters the soft switching technique has a loss of efficiency which is unacceptable, but at power levels above 50 W the technique can offer some reduced harmonic benefits without significantly

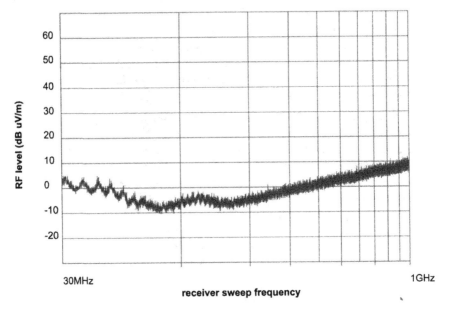

Figure 5.19

Radiated emissions from a toroidal cord DC–DC converter

affecting overall converter efficiency. At even higher power ratings (>200 W) zero crossing schemes can offer similar benefits to power conversion circuits as they offer to motor control circuits, with low conducted noise due to no direct line to ground connections.

Linear regulated DC–DC converters may offer lower noise on the output channels but give very little benefit to the input side and do not provide any advantage over unregulated converters with regards to their EMC performance.

Preference for parts with toroidal or other closed loop magnetic cores and self-resonant types minimise radiated emissions and transformer core ringing. Soft switching types can offer low conducted high frequency spectra, but changing load conditions may upset the switching and create transformer bounce problems.

Conducted interference via supply lines and ground is the most common source of EMI problems, hence the benefits of using a DC–DC converter can outweigh the problems from their switching circuit as they provide a noise barrier between circuits and supply lines. Often adding a fixed frequency DC–DC converter can reduce the system filtering requirements significantly, as only the one switching frequency would be filtered per circuit and hence reduce overall filtering cost.

CHAPTER 6

PRINTED CIRCUIT BOARDS

Design and layout of a printed circuit board (PCB) for electromagnetic compatibility (EMC) considerations is probably the most cost-effective measure it is possible to take in the quest for EMC compliance. It is the most cost-effective as it requires no additional components, just the knowledge and experience of EMC layout methods and time spent in applying them. It may also lead to the reduction in filtering requirements, including the number of filters, hence the application of correct PCB layout for EMC could even reduce the cost of compliance. Critical errors in layout causing an EMC problem may not be resolved by the application of additional filters, therefore relaying the PCB may be the only solution and getting it right first time will always offer the lowest cost approach.

There is no single rule for minimising EMC at the PCB level, it requires the application of many design constraints, most of which are general good design practices. As with most other EMC design rules, the basic principle is to minimise bandwidth, reduce cross-talk and produce quiet signals and systems.

It is probable that not all the ideas postulated here can be applied simultaneously to a given circuit, it is up to the designer or layout engineer to decide which are most applicable. As experience is gained it will become less time consuming and more a matter of course for designers; however, at the moment it appears as an extra design burden. It is not the intention of EMC regulations to prevent certain circuits from ever being produced, it just feels that way. It is worth noting that if a 64-bit 166 MHz computer board can be produced that is EMC compliant, most other circuits should be easier.

6.1 PCB Terminology

It is worth considering the terminology for each possible layer within a PCB so that when reading further on we are all 'singing from the same hymn sheet'. When through hole was the dominant component packaging method life was simpler as there were two easily identified layers; the component layer on which the parts were sited and the tracking layer on the opposite side which contained the interconnect. With surface mount not only does the component side contain the interconnect, but both external surfaces can have components placed on them for maximum packing density.

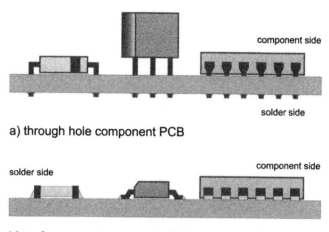

a) through hole component PCB

b) surface mount component PCB

Figure 6.1
Mounting options for PCBs

Component side: any external surface which has the component body sited. For through hole components this is the opposite side to the soldered side. With surface mount both external surfaces could be component sides. The double component sided surface mount PCB is popular in mobile equipment but in most other applications it is still more common to place components on one side only. The component side for surface mount is also the solder side.

Solder side: the side of the PCB to which solder is applied. As noted above this is opposite to the component side for through hole parts and the same as the component side for surface mount.

Buried layer: on a multiple layer PCB, any layer contained within the PCB only (i.e. not on the surface) is called a buried layer.

Tracking layer: this is any layer which contains interconnecting tracks. Tracking can be on the component side, solder side or within the structure of the PCB on boards containing multiple layers. On most PCB designs which contain more than one layer the majority of these layers are tracking.

Plated through holes: vertically interconnecting holes which go between outer surfaces and are metallised (plated) through the hole. These holes are used for through hole components and where the interconnect tracking is on both external surfaces. Plated through holes give very good mechanical attachment to through hole components and are still in common use with connectors and sockets even on boards which otherwise feature all surface mount parts.

Vias: these are interconnections between layers. Vias that go from surface to surface can appear like plated through holes. The function of a via is interconnection not component attachment, hence often vias are much smaller than plated through holes. Vias can also be buried (sometimes known as blind vias), where the via interconnects two tracks one or both of which are buried within the PCB structure.

Ground plane: a continuous plane of copper within a PCB connected to the circuit ground. It is unlikely that the plane can be completely continuous as usually there are through connections such as plated through holes and vias which create small discontinuities within the plane. On some surface mount designs with a simple interconnect pattern it is possible to create a completely contiguous plane on the PCB. The ground plane is usually contained within the PCB and is not placed on a surface side.

Power plane: as above for ground plane, only these planes carry the supply rails for the circuit. It is worth remembering that circuits can require multiple supply rails for their correct operation, hence multiple power planes or a split single plane may be required for functionality.

Laminate: this is the insulating material, usually of a fibreglass resin compound, that the tracking layer is placed on and separates individual tracking layers from each other. It is the laminate that gives the PCB interconnects a different electrical transmission characteristic than standard insulated wires. Laminate layers are also the rigid layers within a PCB that provide the mechanical strength of the finished board.

Prepreg: these are also insulating layers. Prepreg is an abbreviation for pre-impregnated and is sometimes referred to as 'filler'. The prepreg is thinner than the laminate and is a glass cloth filled with epoxy resin. The prepreg is intended to form a layer on top of buried tracking to produce a flat surface on which to add the next laminate layer and bond the laminates together. Prepreg usually has similar electrical properties to the laminate but is a softer compound. Tracking can be made on to prepreg, this type of construction is known as a foil build.

Moat: this is an area which has no tracking or plane through it. A vertical cross-section of a moat area should consist solely of laminate and prepreg layers.

Hopefully none of the above is new to the reader. The main confusion that occurs is the difference between vias and plated through holes. On surface mount boards there may be no through holes and no direct interconnect between outer surfaces, except via buried layers. With through hole boards the component holes make perfectly reasonable via interconnects and save adding to the board complexity with additional vias. Some PCB manufacturers do not always discriminate between vias and component plated through holes, preferring to call them all plated through holes.

6.2 Construction of a PCB

The level of PCB complexity available to the designer will to some extent determine the amount of good EMC design practices that they can put into effect. As an example, if the product cost allows only a single-sided PCB to be used, then ground and power planes are immediately excluded from consideration. Even with a single-sided PCB there are still some good and bad design practices that can influence the EMC of the finished circuit.

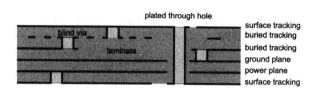

Figure 6.2

6 Layer PCB cross-section

A PCB is constructed from a series of laminate, tracking and prepreg layers in a vertical stack. Within the stack is often a series of drilled holes, plated forming vias between tracking or plated through holes between surfaces. The way these layers are stacked is dependent on the number of layers available for the design. The most popular layer structures are two, four and six layers. The actual number is almost unlimited but the highest in common use is 12 layers. Single-sided (i.e. one layer) boards do appear in very low-cost applications; however, in the majority of modern circuits the limited interconnect possibilities of a single layer are too restrictive.

The process of PCB manufacture has tended to dictate existing common practices for PCB stack structure. The usual advice is to use a balanced stack, that is symmetrical with respect to laminate and prepreg spacings. Symmetrical structure from the inner centre out is also a common theme using equal thicknesses of laminate and prepreg with an even number of tracking layers and ground and supply planes.

There are economical reasons for the use of even layers. The cost of three and four layers is almost equal, the benefit of an additional layer can be costed against a possible increase in packing density (i.e. reduced total board size), hence reduced cost per PCB in volume. In a stack, as the laminate can be metallised on both sides, it is a waste of a laminate layer not to put a pattern or plane on to both sides, hence the PCB stack is going to consist of an even number of metallised layers.

The structure of these layers can be dictated by the user, but there are common themes which are worth considering. The surface layers are almost always tracking layers and rarely contain a plane, this makes sense from a circuit manufacturing viewpoint as it can be impossible to test or debug buried tracking. In four-layer boards the outer layers are the tracking and the inner layers often the power and ground planes. Where a third layer is required for interconnect on a four-layer board the power is sometimes given a routeing pattern within the buried layer and signal tracking added to this layer, but the ground plane is retained. Where power planes are present it is common to have these adjacent to a ground plane and to have tracking on the outside of these, tracking is rarely placed between ground and supply planes.

Multiple ground planes in high layer stack boards are common, but multiple power planes are not used frequently unless required for functionally (e.g. multiple supply rails such as 5 V and 12 V or ±15 V). Where there is more than one power plane, the planes should be separated from each other by a ground plane or on the same plane and be non-overlapping.

6.3 PCB Design Parameters

The design parameters considered important for the EMC aspects of a PCB design are mainly concerned with the mechanical dimensions of the tracking and layers as well as the dielectric properties of the laminate. The values given in Appendix B are standard for a large number of PCB suppliers but you should consult your own supplier to get the most accurate data. Although the absolute parameters used are going to depend on the individual design there are a few considerations that can be applied across all designs.

Cross-talk between tracks can be reduced by increasing the separation between tracking as this is dominated by capacitive coupling. Track impedance can be controlled by tracking above a ground plane (see section on high speed PCB design). The capacitance between power and ground plane can be maximised by using prepreg between the two (foil build), hence reducing the distance between the 'plates' and forming a distributed capacitor. Try and keep the most sensitive and highest frequency tracks away from any power plane as the power supply line tends to contain a relatively high noise content. Minimise impedance on ground and supply rails by using plane layers when possible, where planes are not available use wide tracking. As individual design considerations are discussed later some of these ideas will be reiterated.

If considering a design using signal speeds in excess of 5 MHz, or using devices with rise times of less than 5 ns, then a multilayer board should be planned from the outset. Although single- and double-sided boards are mentioned in the text below, their use is diminishing rapidly and the cost difference between two and four layers is closing. If in doubt at the start of a design on the complexity to allow for in the PCB stack, a guide can be taken from the number of components to be used and the rule of thumb of one tracking layer per five integrated circuits (ICs) or 20 discrete components. This rule is very general and varies so much by design that its usefulness is questionable, but it is better than nothing and can be used at the concept stage to estimate PCB costs.

Tracks on a PCB add inductance, resistance and capacitance to the circuit. The amount of inductance is relatively constant across substrate types and depends on length of track. The inductance per unit length of copper track is similar to that for a component lead, 1 nH/mm. Resistance of a track depends on the cross-sectional area of the track as well as the length, hence values are usually quoted (when they are available) in resistance per square for each weight of copper (see Appendix B). The most popular copper weight, 1 oz, gives a typical value of 0.49 mΩ/□.

Capacitance is calculated using the parallel plate equation:

$$C = \varepsilon_o \varepsilon_r \frac{A}{h}$$

6.1

Where ε_o is the dielectric constant of free space (ε_o = 8.854 pF/m), ε_r is the relative dielectric of the substrate (ε_r = 4.7 for FR4), A is the coverage area and h is the distance between tracks. Therefore a 1 oz copper track, 0.5 mm (0.020") wide, 20 mm (0.8") long over a ground plane on a 0.25 mm (0.010") thick FR4 laminate would exhibit a resistance of 9.8 mΩ, an inductance of 20 nH and a capacitive coupling to ground of 1.66 pF. These values may all seem like low values and even negligible compared with some component parasitics, but there will be many tracks and these will add up to track parasitic elements which equal or exceed several of the component parasitic effects. And, of course, these parasitic effects are under the designers control unlike the component parasitic parameters.

There will, of course, be other design constraints from production or marketing which can restrict some of the options. The restrictions may not be just cost, higher layer stacks create thicker finished boards, some applications such as in handheld equipment and PCMCIA cards may have a severe restriction on board thickness. Similarly, the rigidity of the PCB may be critical for efficient production handling, hence more laminate layers may have to be added to stiffen the product for processing in the manufacture of the finished circuit.

The design constraints for EMC usually come low down on a list of design restrictions but are just as important, if not more so. At the end of the day they can prevent a product reaching the market just as much as the other design constraints. However, few of the other constraints can result in prosecution for failure to comply, which can happen in Europe if a product fails to comply with the EC EMC regulations.

6.4 PCB Layout for EMC

The ideas given are presented in roughly order of effectiveness; therefore, try and apply the first three (segmentation, decoupling and grounding) in some form in all circuit layouts, then as many of the others as is feasible. As mentioned previously, the level of PCB complexity will restrict how many of the layout suggestions given can be implemented.

There may be situations where the only solution to an EMC problem is an increase in PCB complexity to implement more of these design rules. This may prove less expensive than options such as shielding and filtering. It is sometimes difficult to cost out the options when the effectiveness cannot be assessed until the problem is seen and the circuit has been built, or even worse, the end product has failed EMC compliance testing.

6.4.1 Segmentation

Segmentation is also known by the terms quarantining and functional separation. The idea of this principle is to reduce the coupling between circuits by basic physical separation. The actual amount of separation is difficult to specify for all applications and will depend on the wavelength of the signals in each section (one-quarter wavelength gaps being a minimum). As a basic guide, a gap between circuits of approximately 5 mm all around is usually adequate.

Segmentation of circuits is usually performed by using a moated area around each circuit or functional block. Hence there will need to be some patterning of any ground and power planes used. The reason for patterning the ground and supply rails is to prevent a power surge or noise voltage on one circuit block (which may be able to handle the event) being returned via the ground on another circuit block (which cannot tolerate such an event). The ground connections and supply rails will all no doubt meet at the power input to the PCB, but by separating, the loops for supply and ground return are controlled for each circuit.

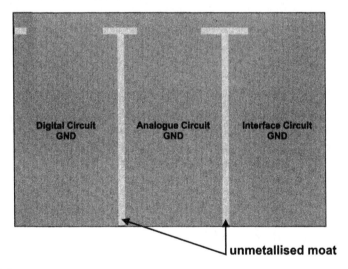

Digital Circuit GND   Analogue Circuit GND   Interface Circuit GND

unmetallised moat

Figure 6.3

Moated ground plane

Segmentation also makes filtering of each circuit block on a PCB simple and reasonably cost effective. For example imagine a PCB has 10 functional blocks using 500 mA each, the total board supply is hence 5 A. A filter for a 5 A DC supply usually contains large inductors, which not only occupy a lot of board area, but are often difficult to place (see section on passive components). Using ten 500 mA filters, the inductors are smaller and each circuit is filtered not only from the power supply unit (PSU) noise, but from each other circuit's noise content. Having segmented circuits makes this filtering option easier. In reality it is unlikely that all circuits require equal current, hence with individual filters the inductor–capacitor combinations can be optimised for each circuit supply requirement and for an appropriate cut-off frequency, some segments may even be unfiltered. The cost of lower current inductors may not be quite a factor of 10 less than the high current version, but the smaller, lower current rated inductors are definitely lower cost and more suitable to automated PCB assembly methods.

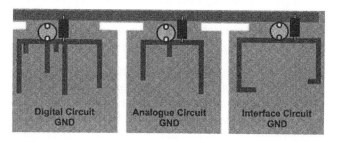

Figure 6.4

Filtered supply rails

Circuits should be segmented by function and/or speed of operation. High speed digital circuits tend to have high instantaneous current demands on clock edges (especially synchronous digital circuits where all the gates switch at the same time, almost). High speed circuits should be placed closer to the PSU inlet than slower circuits such as analogue functions and interface circuits. The placement of high speed circuits close to the PSU reduces the surge demands being observed on the supplies to the other circuit blocks (filtering as mentioned above and adequate

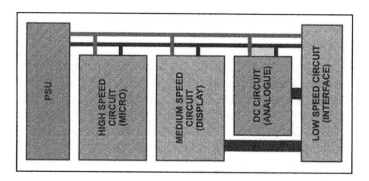

Figure 6.5

Separated function blocks

decoupling also help). It should be borne in mind that it is not the absolute power demand that causes EMC problems within a system, but transients in the power demand.

Circuits which will interface with the outside world or with other PCBs within the end system must be near the PCB edge, there can be no excuse for trailing wires across a PCB within a system. There may be some circuits on a PCB that are known to be noisy, or are intended to handle 'dirty' signals from off-board systems. Filtering may also be required at these circuit inputs, hence within the circuit block a secondary segmentation may be required to handle the off-board signal filters at the PCB interface. A separate moated ground plan for interface circuits would be another good EMC measure, especially if there is a safety ground which could be referenced for ESD and transient suppression circuits directly at the interface socket.

If shielding is ultimately required on certain circuits, by using segmentation at the outset it is easier to shield individual circuit areas rather than the whole PCB. For the purpose of shielding a ground contact will probably need to be available on the PCB surface around the problem circuit. It may therefore be worth bearing this in mind for particularly sensitive circuits and for very high-speed circuits. If not needed the shield can be left off and the only penalty may be a few additional millimetres of gap between circuits, plus the additional ground tracking (see guard ring, covered later in grounding techniques). The provision for shielding of individual circuit segments will not only make the shielding more effective by identifying the critical circuits, but will also minimise the shielding costs as the cost of a shield is primarily governed by the size of the shield required.

6.4.2 Decouple Local Supplies and ICs

The most common way circuits interfere is through their common power supply rails. Maintaining separate decoupling to each circuit area reduces the ability of noise on one circuit's supply rail getting through to another. If space and cost permits filter each supply function on the DC rails as suggested above, a simple LC filter is usually adequate. This filter can remove the need for some mains filtering as it stops noise getting back to the mains supply, as well as reducing cross-talk between circuits. Not forgetting also that DC capacitors and inductors are much smaller and cheaper than their mains counterparts.

Each individual PCB should have a large bypass capacitor (referred to as bulk or reservoir capacitor) at the supply input to the PCB. Bypass and decoupling capacitors are usually not of the same value or construction (see section on capacitors). It would be expected that the bypass capacitor is typically between 10 and 100 µF per PCB and often either an aluminium electrolytic or tantalum type. The bypass capacitor acts as a low frequency ripple filter and potential reserve supply for sudden board demands (e.g. switching operations). The capacitance reservoir reduces localised power demands causing the main supply line to 'dip' by providing current surges locally. The bypass capacitor should be located at the point at which the power enters the PCB, either at the socket or wire terminals connecting the PCB to the system PSU.

Each and every individual IC should have a decoupling capacitor of a few nF, usually of ceramic construction. Decoupling capacitors provide a high frequency, low impedance route to ground for IC switching noise as well as a small reserve for transient power demands. Ceramic is a high frequency dielectric and is the best material for this application (see section on decoupling capacitors for most suitable values).

The decoupling capacitors should always be located close to the IC and closest to the supply pin (usually the Vcc pin). This reduces the tracking impedance to the capacitor, hence maintains the effectiveness of the decoupling capacitor at high frequencies and with fast edge rates. Often the effectiveness of using a decoupling capacitor is lost due to either excessive lead length on through hole parts or excessive tracking on surface mount types. It is more important to have the decoupling capacitor close to the supply lines of the IC it is supposed to decouple, than to the circuit supply inlet.

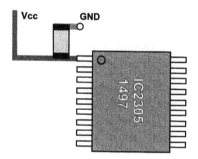

Figure 6.6

Locate decoupling capacitor at IC supply pins

One of the main circuit causes of interference in a digital or mixed signal system is the digital system clock. One way of minimising its break through on the supply rail is to give the clock generating circuit its own decoupling capacitor and if possible a small bypass capacitor. Again these need locating close to the clock circuit and

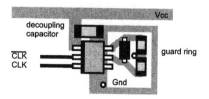

Figure 6.7

Decoupled clock circuit with guard ring

buffers. This whole circuit may benefit from segmenting on the power plane only, the clock and digital grounds should not be segmented. Even filtering may be beneficial on the supply from the digital power supply rail to the local clock circuit. The value of decoupling capacitor will depend to some extent on the clock rate and low values should be used. With clocks up to about 30 MHz a 4.7 nF decoupling capacitor in parallel with a 1 μF reservoir can significantly reduce supply ripple due to the clock and its driver circuit.

6.4.3 Grounding Techniques

This section ties in closely with decoupling and segmentation, but its effectiveness and the techniques available depend very much on the PCB complexity available. The ideal would be to have a multi-layer PCB in which ground and power planes can be defined as separate layers. This can be too expensive for small circuits and there are many single- and double-sided boards that will still need some grounding techniques applied.

The main objective of grounding techniques is to minimise the ground impedance and the size of any potential ground loops from a circuit back to the supply. Note this is not just minimising resistance, at the frequencies of interest for EMC it is inductive reactance of the tracking that usually dominates the impedance characteristic.

One grounding technique that can be used on any complexity of PCB is a guard ring. This is a ground connected track that does not carry a return current for the circuit under normal operation. Its purpose is mainly as a return source for radio-frequency (RF) current that is radiating out of, or incident to, the PCB. It is usually tracked around the outer edge of a PCB, or segments within a PCB, and often around connectors and input–output circuits. If a separate safety earth is being used the guard ring can be connected to this rather than system ground and safety or electrostatic discharge (ESD) devices may sink their current via this track. It is possible to use a guard ring to reduce the segmentation spacing rules as the guard ring can act as a field fringing sink. A guard ring can also be placed around the edge of a power plane as well as tracking layers, this reduces field fringing from uncoupled fields at the edges of a PCB.

On a single-sided PCB, a grounding strategy can still be implemented. The first consideration would be to plan for a wide ground track covering as much of the PCB as possible. Do not attempt a ground plane and then etch out the plane for tracking. This can actually cause more problems than it solves as it may leave unconnected metallised areas within the PCB that can reflect signals through the board or act as receivers and inject capacitively into nearby tracks. A preference for a star arrangement of connecting ground and power should be attempted, but with only a single layer tracking this may be difficult. Definitely apply the segmentation rule of fastest circuits closer to the PSU input to the PCB. Using inductor–capacitor filters at the input to each circuit from a daisy-chained power rail could in fact help with the limited available tracking as the inductors can be used from the supply rail as bridging components. A guard rail can be placed around the edge of the PCB

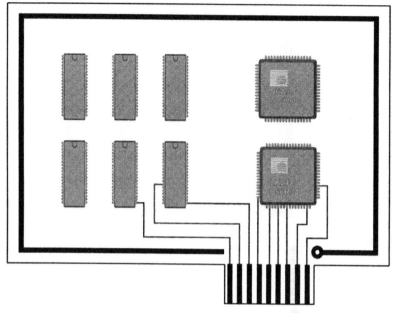

Figure 6.8

Guard ring on component side

connecting to the ground at the input to the PCB only, even on a single-sided board this helps reduce field fringing at the board edge and if a shield becomes necessary leaving the guard ring as uncovered track gives a suitable place to attach the shield.

With a double-sided PCB a ground grid (ground matrix) arrangement should be attempted on digital circuit sections. A ground grid forms a series of box sections on the PCB, which not only reduces the ground impedance but also the size of potential ground loops and signal return loops. A ground plate beneath each IC on the component side will also help even if a full grid cannot be implemented, decoupling capacitors can be tied directly to the IC supply line using this plate. A thick track for the ground grid would be preferred to maintain a low ground impedance, but with high pin count surface mount components this is not always possible. A thin track completing the grid is better than no track as even though this is not a particularly low impedance solution, it still minimises loop areas for both ground currents and signal return paths.

For ground grids to be truly effective at minimising signal loops a similar pattern for the supply should be attempted, mirroring where possible the ground paths. The supply does not have to completely grid in the same way the ground does and comb or star supply arrangements can be very effective coupled with a complete ground grid (comb patterning should not be used on grounding schemes).

With any tracked ground and supply rails, try to track the power rails to mirror the ground, this will increase the effective distributed supply capacitance and minimises

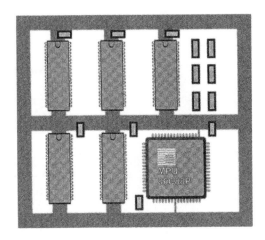

Figure 6.9

Ground grid

the supply-ground loop. Having the supply track slightly narrower than the ground also helps with supply field fringing and reduces cross-talk from the supply rail to nearby signal tracks.

On a multi-layer PCB the ground and power planes should be planned first and usually close together. If one of the supply planes has to be sacrificed for tracking it should always be a power plane, the ground plane should be maintained wherever possible.

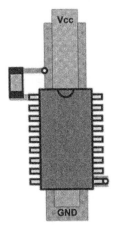

Figure 6.10

Mirror power track over ground track

Many EMC problems encountered with single- and double-sided PCB designs have been solved by simply increasing the PCB stack and including a ground plane. The preferred stack would have ground and power planes separated by a prepreg layer (foil build) or thin laminate, with a thick laminate between power and tracking and a thin laminate between ground and tracking. Using a thin layer between the power and ground planes minimises the distance between these hence maximises the effective capacitance. A PCB capacitor constructed in this manner has a very high-frequency response (high self-resonant frequency) and low series inductance. This construction works as a low value distributed capacitance rather than a bulk single point capacitor. The level of high frequency performance in a distributed plane capacitor cannot usually be found in a discrete component and can be provided by the PCB without adding to the component cost.

Fast circuit signals and sensitive analogue lines should be tracked on the layers closest to the ground plane. The tracking closest to the power plane should be reserved for bias tracks and slower circuits. Using a thicker laminate between power and tracks reduces the capacitive coupling to any signals on this tracking layer, hence reduces the supply cross-talk to these signals. A thin laminate between ground and tracking adds capacitive coupling to the tracks and can allow impedance controlled tracking to be incorporated, but for long track sections or fast pulses this coupling can slow the signal (see section on high speed PCB design).

Again if insufficient layers are available for a separate ground plane, then with sensitive and high frequency signals there should be a ground track beneath the tracking. Producing a ground mirror for these signals ensures minimum loop area, hence minimises radiated signals. For differential signalling the complementary track pairs should be made to mirror each other.

A separate safety ground as either a plane or a guard track is particularly useful where signals enter and exit the system. Often the safety ground cannot be used over a complete PCB plane due to leakage current specifications being exceeded by capacitive coupling effects. A low value decoupling capacitor between signal and safety ground close to the off-board signal connector provides a high frequency

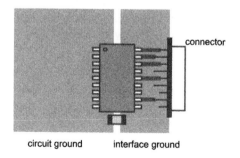

circuit ground   interface ground

Figure 6.11

Interface circuit ground

current ground link between system and safety references. Capacitive coupling between analogue and digital grounds close to any signal interface (e.g. ADC/DAC) should also be planned to bridge the moat region (the capacitor can always be omitted if not required), again low values of capacitive coupling is required (<1 nF).

PCB structure with many layers and a ground plane can implement several of these rules, including a surface ground grid on digital sections with a buried ground plane and even multiple ground planes. In instances where there are several ground circuits or configurations these need to be interconnected to maintain a low impedance and short return loops. Ground stitching is a technique usually referred to placing multiple vias between ground areas at regular intervals. This can be used with guard rings and large grounded areas on PCBs, which are results of copper fill. If the chassis of the system is also grounded using plated through holes for the stitching points and further connecting these to the chassis can produce very quiet PCB designs even with high-speed digital systems.

Copper fill has been a common technique with some analogue circuits, this technique introduces areas of copper on the surface of a PCB which carry no signals and which should be grounded. Although this has the potential to produce a good overall shield to reduce field fringing and improve decoupling, the copper areas can be left unconnected in circumstances and this can produce induced electromagnetic interference (EMI) problems. This technique is not suitable for digital circuits as it can create differences in signal skew and propagation delay between tracks that lead to functional failures. Consequently, copper fill is not a particularly popular scheme in modern designs and should be used with care on analogue circuits only or avoided altogether.

6.4.4 Order of Layout

Layout of the PCB should be done with a certain order of tracks. PCB layout is usually performed on a computer and it is often left to an auto-router to place tracks when using a CAD system. Unfortunately, few of these CAD systems have the intelligence required to consider the implications of how the routeing may affect the EMC of the end system. Indeed, few will have the necessary links to a simulator or netlist to know which tracks are high speed or which are power rails (see Chapter 8 for discussion of CAD programs for EMC). Consequently, some of the more sensitive and fastest tracks may require a manual routeing before the auto-router can be left to finish off the bias and slow signal tracking.

There are few PCB layout programs that can handle the grounding techniques mentioned above, particularly guard rings and segmentation. EMC considerations are therefore going to reduce the reliance on software in the near future as far as routeing and placement are concerned, and manual layout will be required at least for fast and sensitive signal lines.

If multi-layers or ground and power planes are not available, then ground tracking needs consideration before any other tracks. Ground tracks need to be wide and cover as large an area as possible. The highest frequency tracks then need to be laid out, these should be the shortest tracks on the PCB, hence laying them first gives them priority and there should be little other tracking to route around. Sensitive analogue signal tracks need laying down next, then other signal and finally biasing component tracks.

In digital systems the clock or oscillator is the track that should be placed immediately after the power rails. The clock is a particularly problematic track as this usually goes to just about every IC in a synchronous system, hence even when laid first may still be a long track. Additional guard rings may be required for the clock if this is a particularly long section of track to reduce fringing and minimise loop area. The lower order address lines between processors and off-chip memory are usually the next fastest signals after the clock, then data lines.

Digital system tracks should be the same length between processors and each target circuit to minimise skew. Clock lines usually contain the highest frequency content so are the most liable to generate emissions, hence these need to be as short as possible and the clock circuit should be close to or even on-board the fastest operating device (usually the microprocessor). Try and maintain all clock signals to one layer, changing layers changes the characteristic impedance and propagation delay for the signal. By laying down these fastest signals first not only are the track lengths minimised (as there are no other tracks to avoid) but also the number of vias present on these signal lines is minimised.

Another order issue within multilayer boards is the position of each layer within the PCB stack. It is always preferable to have the power and ground plane close (as

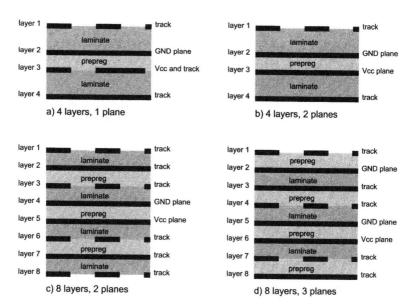

Figure 6.12

Examples of four- and eight-layer track assignments

mentioned previously). The ground should be between the power plane and fastest and most sensitive signal tracks. This may imply having a power plane on an outer layer, which is not always possible due to manufacturing requirements, but could be the quietest solution for a surface mount board with a single component side.

Clock and other high-speed signals should be routed on a layer close to the ground plane. The highest speed signals should track on only one or two layers and these should be adjacent layers. High speed signals should not cross through the ground or power planes if avoidable, this can be achieved with single component sided boards or using multiple layers above or between ground planes. If components are to be placed on both sides and high speed signals tracked through from surface to surface, a second ground plane is probably necessary with the power plane sandwiched between them. Alternatively, a ground mirror track for these high speed signals may have to be patterned between the high speed track and power plane. This mirror track should be three times the width of the signal track to minimise field fringing.

6.4.5 Other Tracking Issues

Where the circuit requires more than one power supply (e.g. ±12 V, or 5 V and 12 V) then these should be patterned into a single layer. If sufficient layers are available to give each supply its own plane then these should be on either side of a ground plane.

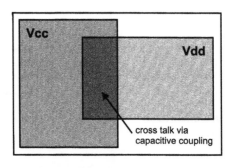

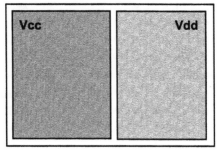

Figure 6.13
Power plane layout

Different supply rails should not be overlapped unless separated by a ground plane, otherwise there is a large potential for coupling, which can upset the system and is difficult to filter. Any overlap of the power planes is a potential cross-talk coupling point between the rails, hence a noisy device attached to one of the supplies is given the opportunity to interfere with any device coupled to the alternative supply rail. A moat separation of about 5 mm should be given between power plane patterns.

To reduce field fringing from the edge of power planes, either single supply or multiple rails, it would also be recommended that the supply plane is slightly inset from the ground plane. An inset of 5 mm gives approximately 60% reduction in field fringing from the edge of the plane.

Where there are two or more tracking layers above a single ground plane cross-talk is minimised by running the tracks orthogonal to one another. Having orthogonal tracks not only helps reduce cross-talk by minimising overlap area, but eases routeing by defining a directional pattern for each layer. Cross-talk is usually produced due to capacitive coupling, so an increase in laminate thickness may help, but this will reduce the effective signal to ground decoupling and change the characteristic track impedance and capacitance.

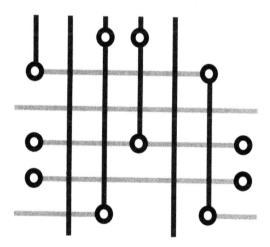

Figure 6.14

Track alternate layers orthogonally

Orthogonal tracking is best suited to lower speed signals, for high-speed signals the discontinuity in the impedance profile of the track, caused by a via, can create reflections and signal loss. The fastest signals should be laid on a single layer and vias and layer changes minimised. A via can introduce between 1 and 4 nH of inductance to a track (depending on via height), with a small capacitive content (0.3–0.8 pF) as the via is perpendicular to any ground plane or ground tracking.

Consequently, vias should be minimised and used sparingly, where unavoidable on high-speed lines the number of vias should be matched on parallel signal lines (i.e. address and data busses).

Where tracks are cornered on a layer, a sharp right angle can produce a field concentration at the inner edge. Although these fields are not strong enough to fail EMC emissions testing they can cause noise injection into other tracks on the PCB. Therefore all corners should be rounded if possible or mitred (angled) by 45°. The radius of bend should be at least twice the track width, or the mitre should contain at least one track width in length along the angled section. Rounded cornering is generally better but not as frequently supported as mitred tracking in PCB CAD packages.

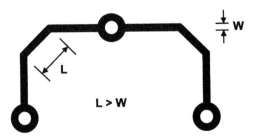

Figure 6.15

Mitre track corners

Right-angled tracking also creates a small discontinuity in the resistance profile of the track (a sharp corner is equivalent to 0.6 of a square of straight track). Although unlikely to cause a reflection in the signal it can make accurate calculation of track resistance awkward.

Avoid producing stubs with tracks carrying high frequency and sensitive signals (i.e. low voltage). Stubs produce reflections as well as potentially adding a wavelength divisible aerials to the circuit. It is not always simple to predict the frequency components in a signal from the simple equations already quoted. So although a stub length may compute to be a non-quarter wavelength integer of any known signal in the system, incident radiation may resonate on a stub.

Another tracking scheme to be avoided is a star signal arrangement where signal tracking radiates from a single point. Although star arrangements are suitable for bonding ground rails from multiple PCBs, with a signal track this introduces multiple stubs. A radiating signal arrangement is usually the shortest tracking and causes minimum delays from source to all receivers, but the multiple reflections and potential for radiated interference create more problems for the circuit's EMI performance.

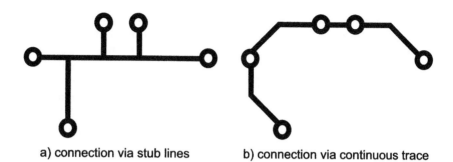

a) connection via stub lines   b) connection via continuous trace

Figure 6.16

Do not use stubs for high frequency tracks

Try to maintain signal tracks at a constant width throughout the length of the track. Varying track width creates changes in track impedance (resistance, inductance and capacitance) and, consequently, the opportunity for reflections and line impedance imbalances. A thin track is less problematic than a varying track. Tracks carrying a bias that is tapped off, and hence can become thinner as they progress, should be funnelled so as to produce a smooth impedance change and remove the possibility of creating a reflection or resonance for interference signals, which are superimposed on the bias.

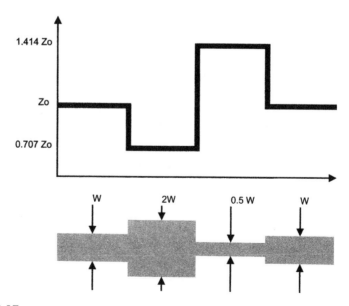

Figure 6.17

Impedance changes with track width variation

Vias and plated through holes in the ground and power planes should be reasonably distributed throughout a layout. A concentration of holes that pass through supply planes produce a localised impedance difference near the holes, hence not only is the area a 'hot spot' of signal activity, but the supply planes are higher impedance at this point and less effective as RF sinks. Hole concentrations cause local tracks to be more susceptible to interference. Where there is a natural concentration of holes (e.g. a through hole connector) the ground plane should surround each hole to minimise local impedance and avoid creating an aperture, this also increases decoupling into the connector or device.

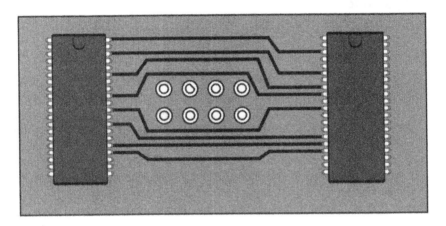

Figure 6.18

Avoid via concentrations

Ground or return signal impedance can also be accidentally increased due to any slits or apertures in the PCB. Apertures should be avoided as they not only increase the ground impedance, but may act as a potential aerial across their length. A series of small holes is preferable to a slit in a ground plane as even though the impedance is slightly increased it is not as problematic as a slit aperture.

Do not deliberately create split apertures (i.e. long holes or wide vias) in power and ground planes. These create an area of non-uniformity within the planes and reduce their effectiveness as shields, as well as locally increasing the impedance of the power and ground planes. This type of feature is usually only required by cabling and cabling should not be brought through a PCB. If the feature is required due to mechanical constraints within the completed system (i.e. physically has to fit over a shaped object) then maintain a ground plane around the aperture and guard rings should be provided on tracking layers.

Do not leave any unconnected metallised patterns. If components have a pad which is not connected (often marked NC on the pin-out or schematic diagram) but soldered down, connect to the ground. If the pin is suspected to be internally

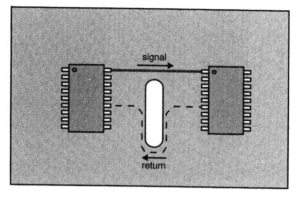

increase in path length due to split aperature

Figure 6.19

Avoid slit apertures in the PCB

connected contact the supplier for confirmation, otherwise route to a convenient point. Unconnected metallisation can act as an aerial and inject into the local circuit capacitively. Often unconnected metallisation is used for thermal dissipation of the device within the package, hence grounding is the best method as the ground plane/track should have the highest amount of metal to act as a heat sink. Always check the data sheet or call the manufacturer if unsure.

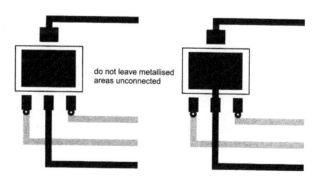

do not leave metallised
areas unconnected

Figure 6.20

Connect all metallised areas

Minimise loop areas by keeping tracks to and from components and sockets close together (signal and returns). This reduces the potential aerial loop size and the capacitive coupling between these tracks minimises bandwidth. With high speed single-ended signals sometimes the ground return may also have to be tracked alongside the signal if the signal does not track over a low impedance ground plane.

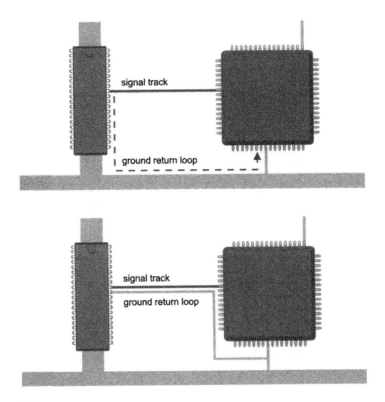

Figure 6.21

Minimise return loop area

Many of the tracking issues may seem trivial, they may appear not to be likely sources of emissions from the system. It is not just emission out of the system that could be the ultimate problem, but emissions within the system causing internal interference or injecting back into the circuits to produce conducted noise. Consider an unconnected metallised area for a small component back plate, a SOT223 transistor for example. The metallised plate will reflect any incident radiated field, and rotate the field through 90°. If the incident radiation failed to interfere the first time it passed through a local component, with its plane rotated the radiation gets a second chance, it may even be rotated to pass through the plane of the PCB itself. Although individually each idea presented here may produce a negligible influence on the EMC of a system, it is the sum of many of these design principles which can reduce both emissions and susceptibility of the completed design.

CHAPTER 7

PCB DESIGN FOR HIGH-SPEED CIRCUITS

The previous layout suggestions are relatively general and can be applied to most circuits to reduce electromagnetic compatibility (EMC) problems, almost regardless of the circuit function. In high-speed systems, where there are many signal lines carrying high-speed signals with high rise and fall times (e.g. clock signals, data busses, address busses) attention needs to be paid to the effect of track and termination impedance using transmission line principles.

The correct application of transmission line principles also requires the application of some mathematics. Whereas the previous suggestions may have seemed intuitive and their application not very scientific, they have in fact been derived from some of the following principles and mathematical derivations. The maths has deliberately been left out of the previous suggestions to make their application more accessible. Unfortunately, the following principles rely on more rigorous rules and formulae that cannot be too generalised if they are to work correctly.

7.1 Controlled Impedance Tracking

In a high-speed digital system it is not sufficient to simply lay all the fast signal tracks 'first', as most signal lines could be classified as 'fast'. As well as keeping these tracks short they must maintain very close impedance matching to each other and their end termination needs to match the line impedance to reduce signal delay and reflection. What this suggests is that at high-signal speeds (typically in excess of 50 MHz) there is going to need to be as much attention paid to the mechanics of layout detail as is given to the circuit design itself. The techniques for layout are similar to those used by designers of microwave circuits, where physical layout and track dimensions are an integral part of the circuit design process.

It is when dealing with high-speed designs and trying to produce impedance controlled tracking that the need for the mechanical printed circuit board (PCB) design data is required (see Appendix B). The track mechanical dimensions are important as well as PCB layer characteristics, particularly track width (w), track thickness (t) and track length (l). These are defined by the designer at the layout stage, although length is dictated as much by component siting and pin-out as by deliberate track design. Consequently, the PCB designer can exhibit full control over these parameters and minimise EMC and potential high-speed functionality problems at a very early stage.

Control of the PCB material and properties needs exercising by the manufacturer of the PCB and you may need to use a different supplier for high-speed PCBs than low-speed boards. In particular, the supplier has to be able to offer controlled impedance boards to ensure that all the design effort put into ensuring the PCB design is optimised for EMC performance is not lost due to poor quality control at the PCB supplier.

There are two forms of controlled impedance tracks (transmission lines) possible on a PCB: the microstrip and the stripline. The microstrip is a track on the surface of the PCB running above a ground plane. The stripline is embedded within the PCB and between two ground planes. Although the text refers to the tracks being over a ground plane, the supply plane also provides a coupling reference for tracking, hence the tracks above supply planes or between supply and ground also utilise the transmission line track principles. Each tracking scheme will be dealt with separately as the equations for dealing with them are different. It is possible to mix the two and certainly not uncommon to have both on a PCB. In fact it would be difficult to have an embedded stripline only without surface microstrip tracking, as any surface tracks would share a common ground plane, even if the surface tracks had not been deliberately designed as high frequency tracks.

The characteristic impedance of a track is a measure of the characteristic inductance (L_o) and capacitance (C_o) per unit length of track;

$$Z_o = \sqrt{\frac{L_o}{C_o}} \qquad\qquad 7.1$$

The characteristic impedance is measured in ohms (Ω) and in itself does not vary with track length. It is this impedance that is to be controlled and matched between synchronous signals if correct high-speed designs are to be realised without EMC problems.

Another parameter which is important to high-speed signals and in transmission line application is the speed of the signal (signal velocity, v_s). The electrical signal travels at a slower rate than in a bare copper wire due to the influence of the dielectric medium:

$$v_s = \frac{c}{\sqrt{\varepsilon_{eff}}} \qquad\qquad 7.2$$

Where c is the speed of the electromagnetic wave in free space (3×10^{-8} m/s) and ε_{eff} is the effective relative dielectric of the substrate. In a high-speed circuit it is not the actual velocity that is important but the propagation delay this reduced velocity produces. The propagation delay (τ_{pd}) is the reciprocal of the velocity but usually expressed in appropriate dimensional terms such as picoseconds per mm (ps/mm) or nanoseconds per foot (ns/ft). (Note: 10 ps/mm = 3.05 ns/ft)

The propagation delay can also be expressed in terms of the characteristic inductance and capacitance of the track:

$$\tau_{pd} = \sqrt{L_o C_o}$$ 7.3

The characteristic inductance and capacitance could therefore be found for any given track from determining impedance and propagation delay. The characteristic inductance varies with the track width, length and to a small extent track thickness, in general wider tracking reduces the characteristic inductance. The characteristic capacitance also varies with track dimensions and with PCB dielectric and layer thickness, wider tracking increases the characteristic capacitance as does a thinner layer build. Although characteristic inductance is rarely used except for some simulation exercises, the characteristic capacitance is an important parameter to determine the loading on driver circuits, and this can in itself limit the design freedom available with certain impedance controlled circuit boards.

General equations for the characteristic inductance and capacitance can be derived from the above characteristic impedance and propagation delay equations:

$$L_o = Z_o \tau_{pd} \text{ and } C_o = \frac{\tau_{pd}}{Z_o}$$ 7.4

These equations can be used where necessary, but caution should be used with the units especially if using mixed imperial and metric terms. For example a track of 50Ω with a propagation delay of 10 ps/mm, will exhibit 0.5 nH/mm characteristic inductance and 0.2 pF/mm characteristic capacitance. These equations are usually easier to use than rearranging the equations derived directly from the impedance that follow and include track dimensions directly, also any anomalies due to dielectric constant need only be accounted for once in the impedance or propagation delay calculations.

7.1.1 Surface Microstrip Tracking

The impedance of a single track over a reference plane is controlled by maintaining a fixed distance to the plane and using constant track width over the length of the track.

Figure 7.1

Surface microstrip track

When a single ground plane is used, either on double sided or multilayer PCB, and the tracking is on the surface the characteristic impedance (Z_o) is given by:

$$Z_o = \frac{87}{\sqrt{\varepsilon_{eff} + 1.41}} \ln \left(\frac{5.98 \ h}{0.8 \ w + t}\right)$$ 7.5

Where h is the distance between the track and ground plane (layer thickness), w is the track width, t is the copper thickness and ε_{eff} is the effective relative dielectric of the substrate.

Some confusion has occurred with the value for effective dielectric constant for microstrip applications. This is due to the fact that the signal field travels partially through air and partially through the laminate dielectric for a surface microstrip track, hence the effective dielectric constant (ε_{eff}) can be calculated from the PCB dielectric constant (ε_r) using the equation:

$$\varepsilon_{eff} = \left(\frac{\varepsilon_r + 1}{2}\right) + \left(\frac{\varepsilon_r - 1}{2}\right)\left(\frac{1}{\sqrt{1 + 10\dfrac{h}{w}}}\right) \qquad 7.6$$

This can be quite cumbersome and a value for FR4 of $\varepsilon_{eff} = 3.5$ is commonly used for surface microstrip tracking. As a first approximation for other materials a divisor of 1.35 could be used for the commonly quoted relative dielectric value. Note this does not affect the value used for the calculation of line capacitance (i.e. $\varepsilon_r = 4.7$ for FR4) as the effective capacitance is contained solely in the dielectric material and is not fringing the air above the track.

It is quite usual to ignore the track thickness term (t) as this is much smaller than the width ($t<<w$) and this simplifies the maths. As an example, consider a 1 mm wide track which is 0.5 mm above the ground plane on an FR4 substrate, the transmission line impedance approximates to 52Ω, hence a similar value of termination would be required to match this on a high-speed digital signal line and the driving circuit should be capable of feeding this value of impedance. This value would also provide a good match for a 50Ω coaxial cable interface track (between driver/receiver circuit and connector).

An additional design constraint with surface microstrip lines is the effect of the solder resist mask. The solder resist adds a layer of dielectric above the tracking and can effectively reduce the impedance by between 1Ω and 3Ω depending on thickness. This effect of solder resist is commonly ignored, but in long lines, low-impedance tracks or matching between different types of track construction its effect may have to be considered.

The speed of the signal is affected by this tracking and will need consideration for matching between circuits such as distributed clocks in a synchronous digital system. The propagation delay (τ_{pd}) of a microstrip track is governed by the dielectric of the substrate via the equation:

$$\tau_{pd} = 3.337 \sqrt{0.475\,\varepsilon_r + 0.68\varepsilon_{eff}} \qquad 7.7$$

The equation as written is in units of picoseconds per mm (ps/mm). This equation is commonly used as an empirical approximation for microstrip tracking rather than the standard velocity equation and effective dielectric constant. There are differences in the results when using the above equation and the theoretical standard velocity equation, which are relatively small. The above equation has become a standard textbook formula and is used often without knowledge of the approximations to the effective dielectric implied ($\varepsilon_{eff} = 0.475\ \varepsilon_r + 0.68$). I believe one of the primary reasons for the use of this approximation is the poor control that is exhibited in the dielectric constant of PCB laminate materials, resulting in simple mathematical approximations of this type becoming *de facto* standards. The above equation will be used here for consistency with other texts; however, the theoretical equation below is believed to be a more accurate formula providing an accurate effective dielectric constant can be determined:

$$\tau_{pd} = 3.337\ \sqrt{\varepsilon_{eff}} \qquad\qquad 7.8$$

It is often the delay between tracks that is more important than absolute signal propagation delay, as the delay between a clock and data signal can affect functionality as well as EMC. For example, on standard FR4 the signal delay is 5.69 ps/mm, imagine we had a microprocessor in a 255 pin pinned grid array (PGA) package, where pin 1 is the master clock (MCLK), and pin 21 is the lowest address line (ADDR0), separated by 40 mm across the package. Ignoring any internal delays within the silicon, a device near pin 1 which requires both clock and this address would see a 228 ps signal edge delay when these pins are supposedly synchronously switched. Although relatively minor in itself we are now beginning to broaden the switching edges, even though the frequency remains constant, we are introducing small delays into the switching edges and hence to the power demand.

The delay may seem minor on a single package; however, over say a 300 mm backplane the signal delay is just over 1.7 ns. As clock speeds exceed 50 MHz in a system this delay is close to the rise time and a significant part of the pulse width, functionality as well as EMC may be compromised. Certainly, real-time systems would have trouble synchronising multiple processing boards over such a large backplane at 50 MHz data rates.

7.1.2 Embedded Microstrip Tracking

A microstrip track can be embedded in the laminate if the PCB has a build on top of the microstrip tracking layer, for bias tracking for instance or to protect the tracking integrity during circuit manufacture. Although relatively uncommon this arrangement is possible. The above equations still hold, but the characteristic impedance is changed as the effective dielectric constant approaches that of the laminate.

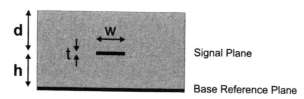

Figure 7.2

Embedded microstrip

If the distance between the microstrip and ground plane (h) is equal to, or less than, the distance between the microstrip and air (d) then the relative dielectric constant of the laminate should be used to calculate the characteristic impedance (i.e. if $h \leq d$ then $\varepsilon_{eff} = \varepsilon_r$).

For a surface layer build which is in between the track to plane distance ($0 < d < h$), the effective dielectric constant can be calculated from the equation:

$$\varepsilon_{eff} = \varepsilon_r[1 - \exp(\frac{-1.55(h + d)}{h})] \qquad 7.9$$

This value can then be substituted into the microstrip impedance equation. If we re-examine our 1 mm wide track 0.5 mm above the ground plane with a prepreg build of 0.25 mm above ($\varepsilon_{eff} = 4.24$), the characteristic impedance is 48Ω, a 4Ω decrease over the surface stripline value. This may seem an insignificant difference and is one of the reasons that many designers use the standard dielectric constant (ε_r) rather than the effective dielectric constant (ε_{eff}), as the differences produced are usually less than 10% of the actual value.

Using the de facto standard equation for propagation delay would indicate that there is no change in signal velocity. This is not true but the changes are relatively small until the dielectric layer above the track is equal to, or greater than, the track to ground separation. At this point the signal velocity should equal that of a stripline as the signal is travelling completely in the PCB dielectric medium.

These equations will be required where there are several layers carrying high-speed signals over a single ground plane. Impedance matching between signals on different layers can be achieved by careful design. If a signal is tracked over several layers the mathematics become quite complex as the characteristic impedance changes between layers, and at vias, hence reflections could occur and this could be one situation where changing the track width becomes necessary to impedance match between tracking layers. Maintaining signals tracks between devices within a single layer will always provide the simplest analysis and usually the quietest solution.

7.1.3 Stripline Tracking

Embedded tracking between two ground planes (or power planes) which are equally spaced from the tracking layer is considered as balanced stripline tracking. This is the

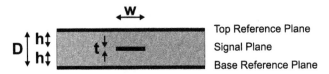

Figure 7.3

Embedded stripline track

quietest tracking method for signals as the field is contained totally within the PCB laminate, although some fringing can occur if the track is close to the PCB edge.

With a track embedded between two equidistant ground planes, the track impedance is given by:

$$Z_o = \frac{60}{\sqrt{\varepsilon_r}} \ln(\frac{1.9D}{0.8\ w + t})$$ 7.10

Where D is the separation of the ground planes ($D = 2h+t$). In using stripline tracks the signal field is contained within the PCB so the standard relative dielectric constant of the laminate can be used (i.e. $\varepsilon_r = 4.7$ for FR4). For example, again consider a 1 mm wide track embedded between two ground planes 1 mm apart (for comparison with the microstrip calculation and assuming $t<<w$). The stripline track characteristic impedance is 24Ω, less than half that of the surface microstrip using a similar build and half of the embedded microstrip construction. Stripline can be equated to parallel embedded microstrips of similar build, hence stripline will offer a lower impedance for a similar build structure than its microstrip equivalent.

The stripline does produce a slightly lower signal velocity than the microstrip, hence an increased signal propagation delay:

$$\tau_{pd} = 3.337\sqrt{\varepsilon_r}$$ 7.11

On standard FR4 the signal delay is 7.23 ps/mm, indicating that microstrip may be better than stripline for the fastest signals with respect to propagation delay. But stripline is the lowest noise tracking scheme from a radiated noise viewpoint as the tracks are fully enclosed on two sides and the impedance is lower.

Considering our previous example of a 255 PGA with 40 mm pin spacing between synchronised signals, the delay is 289 ps between pins at opposite edges of the package. Over a 300 mm backplane the delay is increased to 2.1 ns using stripline compared with the microstrip 1.7 ns. It is physical effects such as this that limit address and data backplane speeds in digital systems to frequencies much lower than the microprocessor clock speed.

The stripline can be further enhanced for signal immunity and radiation by running vias along either side, producing a completed ground shield, the track then approximates to a coaxial cable and the above equations no longer hold. This would produce a further reduction in signal velocity and hence increase in propagation delay. In general, producing coaxial structures within the PCB is not required and adequate noise performance can be achieved by using a suitable transmission line structure and spacing tracks adequately.

7.1.4 Dual Stripline Tracking

Striplines can have multiple tracks on separate layers within the stripline structure, or have single tracks which are offset from each of the ground planes. This changes the equations for characteristic impedance, propagation delay remains unchanged as the effective dielectric constant remains the same. The characteristic impedance becomes:

$$Z_o = \frac{80}{\sqrt{\varepsilon_r}}[1 - \frac{h}{4(h+d+t)}] \ln [\frac{1.9(2h+t)}{0.8\,w+t}]$$ 7.12

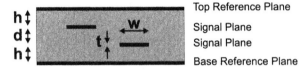

Figure 7.4

Dual stripline tracking

The impedance of a dual stripline using 1 mm wide tracks with 0.5 mm layer builds ($h = d = 0.5$ mm) would be 28Ω (assuming $w >> t$ and $h >> t$).

It would be expected that alternate layers within this type of dual stripline build would be tracked orthogonally (i.e. at right angles to each other). This minimises cross-talk by minimising common capacitance between tracks. With tracking layers equally spaced from the ground planes, the impedance for a track is the same on either layer, hence this would be a suitable method for tracking complex wiring arrangements over two tracking layers with matched impedances.

7.1.5 Cross-over Capacitance

Cross-talk occurs between signal lines which have mutual inductance or capacitive coupling. Mutual inductance within a PCB is extremely small and most track cross-talk is generated due to mutual capacitance. The major capacitive coupling for PCB tracking occurs at cross-over points, where tracking from different layers cross.

The individual cross-over points each have a very small amount of capacitive coupling; however, there may be tens or hundreds of cross-overs, hence the coupling can become significant.

The track-to-track capacitance can be estimated using the equation:

$$C_c = \varepsilon_o \varepsilon_r \frac{(w_1 + 0.8h)\,(w_2 + 0.8h)}{h}$$ 7.13

This differs from the standard parallel plate equation due to the influence of the track close to the actual cross-over creating a slightly modified area value. The equation depends on the distance between the tracks (h) being small and providing the track widths (w_1 and w_2) are at least twice the separation ($w_1 \geq 0.5h$, $w_2 \geq 0.5h$). When

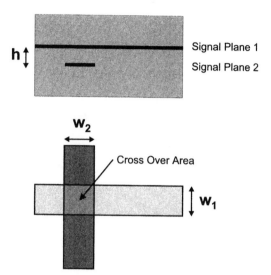

Figure 7.5

Cross-over tracking

using fine tracks or large separations the standard parallel plate equation will suffice as a first approximation.

On standard FR4 with two 1 mm tracks separated by 0.5 mm the single cross-over capacitance is 0.16 pF (ε_o = 8.854 pF/m, ε_r = 4.7). This magnitude of separation is relatively large and generally the separation on a dense PCB will be in the region of 0.2 mm, increasing the coupling capacitance to 0.4 pF per cross-over. Although the value per cross-over is small it can rapidly sum for dense PCBs to several pico-Farads of capacitance which will not only slow signals but will couple many signal lines together.

7.1.6 Termination and Line Length

A transmission line with a characteristic impedance (Z_o) will require resistive termination that matches this line impedance, which allows transmission of wideband signals without degradation. If the trace is unterminated reflections and ringing can occur due to the impedance mismatch. These noise problems due to impedance mismatch can create false triggering and offset switching levels, hence reduce noise margin and create emissions.

Not all lines necessarily require termination; there is a critical length (l_{max}) at which the tracking becomes 'electrically long'. The on-set of an electrically long track occurs when the propagation delay of the tracking (τ_{pd}) becomes a significant part of the signal rise time (τ_r), defined by the Nyquist criteria:

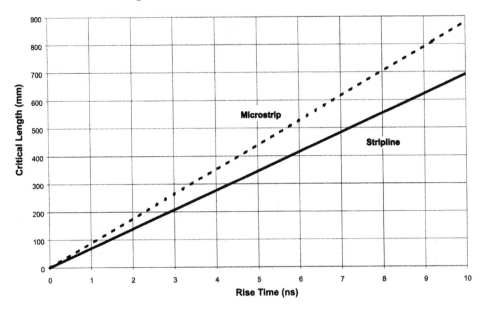

Critical Length of Transmission Line Tracking on FR4 Substrate

Figure 7.6

Critical length for FR4

$$l_{max} = \frac{\tau_r}{2\tau_{pd}}$$

7.14

With knowledge of the rise time and tracking structure (microstrip or stripline) this critical track length can be determined. For example, assuming ALS logic is used with a maximum rise time quoted as 4 ns, using microstrip tracking on FR4 the maximum unterminated line length is 351 and 277 mm for a stripline track. These seem very long and except for backplane design it could be inferred from these results that most tracking will not be electrically long, but problems could be encountered as the specification for rise time is the maximum value. In reality the ALS logic drivers could easily be exhibiting typical rise times of 0.5 ns, which would see microstrip lines of 44 mm or stripline tracks of 35 mm as electrically long. With modern integrated circuit (IC) packages these distances can occur across the package, let alone between different devices.

As IC manufacturers reduce the dimensions of their transistors, the rise times of their devices will inevitably fall, hence in PCB design termination is going to become necessary to maintain upgradability. Last year's ALS logic device may have twice the rise time of this year's, but the specification will remain the same so there will be no way of knowing what minimum rise time the device will exhibit.

7.1.7 Device Loading

So far the discussion has concentrated on the effects of the tracking line only; however, the true line impedance and propagation delay will also be affected by the loading of devices on the line. Even unconnected device pins and sockets add up to 4 pF of capacitance and vias and plated through holes can add up to 0.8 pF of capacitance and up to 4 nH of inductance per feature. Consequently, the true transmission line impedance should include these effects.

It is the capacitive loading (C_D) of the devices and track features that produce the loaded impedance (Z_{LO}) and propagation delay (τ_{Lpd}) of a track.

$$Z_{LO} = \frac{Z_o}{\sqrt{1 + \dfrac{C_D}{C_O}}} \qquad 7.15$$

$$\tau_{Lpd} = \tau_{pd}\sqrt{1 + \frac{C_D}{C_O}} \qquad 7.16$$

It is this loaded propagation delay (τ_{Lpd}) that should be used to determine if a track is a transmission line (i.e. electrically long). This will of course further reduce the effective line length prior to the onset of transmission line effects and slow signal velocity within a system. Consequently, for accurate matching of signal delays and impedance, not only does the track have to be properly designed, but device loading should be matched. Device load matching is usually accounted for by the fact that critical signals generally are between the same devices (e.g. between memory and processors, between interfaces and buffer circuitry), only the clock signals may experience significant loading differences.

As an example consider a clock line driving ten 5 pF gates over a 10 cm balanced 50Ω stripline. The characteristic line capacitance is 14.46 pF, hence the loaded impedance is 23.7Ω and the loaded propagation delay is 15.3 ps/mm, half the impedance and twice the delay of the unloaded line. This is part of the reason high-speed clocks have such stringent fan-out rules and why buffering the clock is carried out both on-chip and between devices.

If possible it would also be recommended that loads are evenly distributed along a transmission line. This can be difficult to achieve as the location of devices which load the line are not defined at the initial stages of layout due to their loading effect but their interaction. Hence by the time an assessment of the loading is made the line length and device positions are often already fixed.

7.1.8 Reflection and Ringing

The main purpose of using transmission line principles for the design of high-speed signal PCB tracks is to enable signals to be transmitted with the minimum amount of noise and interference. The main source of the interference will come from the signal itself, interacting with the line and loading to create ringing and reflection artefacts in the PCB trace. These can be minimised if we know the line and loading impedance (Z_{LO}) and the impedance of the driver source (Z_S) and receiver circuit (Z_R).

Reflections are initially caused by mismatch between the impedance of the receiving circuit and the line, the coefficient of reflection at the load (ρ_R) is given by the ratio of the impedance mismatch.

$$\rho_R = \frac{Z_S - Z_{LO}}{Z_S + Z_{LO}}$$

7.17

The mismatch in load and drive source impedance also causes a reflection, again the source reflection coefficient (ρ_S) being the ratio of impedance mismatch.

$$\rho_s = \frac{Z_S - Z_{LO}}{Z_S + Z_{LO}}$$

7.18

The receiver impedance is often high in many logic devices, hence the receiver reflection coefficient is almost 1 ($\rho_R \approx 1$), this produces a full reflection of the drive voltage back to the driver. Consequently, it is most often the driver to loaded line impedance that is the critical parameter determining ringing and reflections in a transmission line track.

If the driver source impedance is less than the loaded line impedance ($Z_S > Z_{LO}$) then reflections at either end of the line occur and a ringing waveform is seen in the line (the signal is 'bounced' from driver to receiver many times, diminishing in magnitude after each reflection). If the source impedance is less than the loaded line impedance ($Z_S < Z_{LO}$) then the driver experiences difficulty in driving the signal into the track and a staircase or stair-stepping voltage is observed on the line.

The ideal is to have matched driver and receiver impedances; however, this is often difficult, particularly with many devices coupled to the line (e.g. microprocessors,

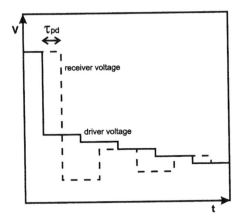

Figure 7.7

Staircase stepping voltage due to impedance mismatch

memory and interface on an address bus). It is due to the multiple number of devices that can receive simultaneously that many of the receiver circuits have such a high impedance. Driver impedances are sometimes given, so these may not be as difficult to determine; however, often the high-to-low and low-to-high transitions may offer difference impedances, hence matching of the driver may in fact not be possible. If impedances are unknown but suspected to be significantly higher at the receiver than the source/drive, termination of the transmission line will be necessary to reduce noise problems and ensure signal integrity.

7.1.9 Line Termination

To minimise reflection and ringing transmission lines may need to be terminated using additional components. The type of termination will depend on the line impedance and the drivers and receivers connected along the line.

Circuits which are driving into a line from a single driver with multiple receivers (such as clock circuits) usually have a lower driver source impedance than the line or load impedance. To match the source to either line or load, a series resistor is placed in line with the drive (Figure 7.8a). The value of resistor is usually quite low (typically 10Ω) and usually matches the driver impedance to the line characteristic impedance.

$$R_t = Z_{LO} - Z_S \qquad\qquad 7.19$$

The resistor is usually placed immediately after the driver output, between the driving circuit and the transmission line. The inclusion of a resistor in series can cause additional signal delays, especially into capacitively loaded receivers, although with most digital circuits having the clock delayed slightly compared with the data can have beneficial results for system integrity.

The pull down resistor (shunt resistor or parallel termination, Figure 7.8b) is a common technique and is easy to implement. Only one resistor per line is required not one per receiver, unless each resistor is a high value so that the total parallel resistance is equal to the characteristic line resistance. Multiple resistors per receiver circuit is a method which creates an even distribution of terminating impedance, but usually the higher component count is a problem and a single resistor per transmission line is most frequently used. The shunt resistor does not create additional propagation delays but does increase the line drive requirement as the termination impedance is reduced. The line driving circuits must be capable of driving the termination impedance as well as the line impedance, which should be matched ($R_t = Z_{LO}$).

Where the increased DC current drive is unacceptable or undesired the termination can be AC coupled by the addition of a single capacitor per line (Figure 7.7c). The capacitor has to have a low value and high operating frequency (typically 100 pF MLCC type) so that at the transition the impedance of the capacitor is low (near zero) and hence the termination impedance is equal to the resistor alone. The DC impedance after transition reverts to the terminating impedance of the receiver

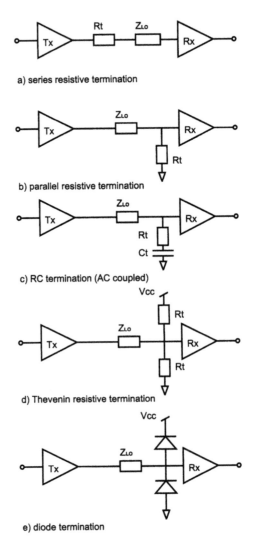

a) series resistive termination

b) parallel resistive termination

c) RC termination (AC coupled)

d) Thevenin resistive termination

e) diode termination

Figure 7.8

Termination schemes

circuits, which is usually high, hence there is little DC current flow after transition. This technique reduces the line overshoot and ringing without significantly affecting the steady-state current requirement of the complete circuit. Again the terminating resistor is chosen to match the characteristic line impedance ($R_t = Z_{LO}$). The additional termination capacitance will add to the loading capacitance of the circuit, slowing the signal and reducing the loaded line impedance.

Circuits which feature different impedance for driving high-to-low and low-to-high transitions can be terminated by a Thevenin termination with both pull-up and pull-down resistors (Figure 7.8d). For example, VME bus systems usually feature a 470Ω pull-down and a 330Ω pull-up resistor. This increases current consumption as whether the line is driven high or low, there will be a current flow through one of the resistors. There is also a DC offset on the line of approximately half the supply rail which may reduce the susceptibility of the circuit if the line is in a non-active drive state.

Overshoot and undershoot can be significantly reduced by use of diode termination (Figure 7.8e), this clips both positive and negative transitions to within a single diode drop (typically 0.6 V) without the increased DC power consumption of the Thevenin termination. The diodes have to be selected to be fast enough to operate at the transition rates of the signals and with sufficiently low capacitance to have negligible effect on the line loading. Diode packs featuring multiple line termination pairs are available to simplify placement of this type of termination on bus lines. Strictly speaking diodes do not terminate the line as they are not matching any impedances; however, they may be an adequate solution if the main EMC or signal integrity problem is overshoot or undershoot (or reflections due to these effects), they also provide electrostatic discharge (ESD) protection of the lines.

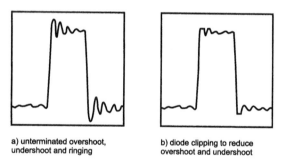

a) unterminated overshoot, undershoot and ringing

b) diode clipping to reduce overshoot and undershoot

Figure 7.9

Diode termination reduces overshoot and ringing

The above termination principles should be applied to differential lines as well as the single-ended transmission schemes shown in Figure 7.8. Differential lines may be individually terminated or cross-linked by resistive or diode schemes.

7.1.10 Which Transmission Line?

Microstrip tracking is more popular than stripline due to its lower cost and greater simplicity. There is also additional test and production benefits in having the tracks accessible at the surface of the PCB that makes microstrip more attractive to many users, but surface tracking also increases the potential for radiated emissions. Microstrip is faster than stripline, however, as noted above for similar dimensions of track microstrip exhibits higher impedance than stripline and is consequently more susceptible to electromagnetic interference.

The stripline method requires at least one additional ground plane, probably several in complex microprocessor designs. The enclosed track construction ensures that both emissions and susceptibility are minimised, particularly useful for long runs such as address bus backplanes. The lower impedance is another beneficial parameter when creating long tracks, but impedance matching to low value lines can increase power consumption (i.e. it takes more power to drive 24 Ω lines than 48 Ω lines at the same data rate).

There are no straightforward rules; it would be recommended that clock lines were distributed via stripline tracking as these are usually the greatest potential noise source in a digital system. Also, having the clock travelling at a slightly slower rate than the signals can help with ensuring that the signal lines have stabilised before the clock signal latches the gates. Having a clock distributed to many circuits can also help in creating the low termination impedance due to the parallel effects of many higher impedance terminations; however, fan-out rules must still be obeyed, regardless of PCB tracking method.

Shorter tracks between local ICs can use microstrip tracking in most cases. It is unlikely that all tracks will be able to be adequately matched for signal delays as many of the larger IC packages can exceed 50 mm between pins on a single device (e.g. 255 pin PGA). Using straight through embedded microstrip tracks could create a time delay between signal edges of over 120 ps on large IC packages. If the system clock frequency exceeds 100 MHz this signal delay approaches 2.5% of the pulse width and begins to become a potential signal integrity problem. As IC manufacturers increase the functionality of silicon devices and pack in more transistors per square millimetre, the size of the packaging seen on PCBs is likely to increase. The majority of these larger devices have all the address lines close together and all the data lines close together, so common sense is being used in the IC design and packaging.

Using serpentine tracks to match the delay for some of the signal tracks is not a common technique but can be implemented when absolute matching of delays is required. Mixing surface and embedded microstrip tracking with stripline is an alternative technique to match the delays between high-density signal tracks across a large IC device package or to a backplane or connector.

7.2 Multilayer Build

The construction of layers within a PCB can help with the EMC performance. In particular, with controlled impedance boards it may be necessary to add additional ground and power planes, and the location of these can be chosen to help reduce emissions and susceptibility.

This at first sounds like a recipe for more layers equals better EMC performance. While this may be generally a truism, it has to be tempered with some design input as simply increasing the layer stack in a PCB alone is not the solution. One problem that can occur and should be considered is the overall PCB thickness. The closer the tracking the greater the potential for cross-talk between signals, the lower the impedance and higher the drive (power) requirement, hence greater immunity but with a higher potential for conducted noise and higher manufacturing cost.

7.2.1 Number of Layers

The number of layers is almost impossible to calculate with any accuracy or certainty. The previous chapter suggested one tracking layer per five ICs, which is a very loose approximation for simple designs. For high-speed systems the tracking density would be expected to be increased, but the number of layers would not be of this order of magnitude, otherwise most digital systems would have over 20 layers within their PCB.

For high density digital systems an approximation can be obtained from the number of address bits (A_n) and the processor speed (f_{CLK} in MHz). Higher bit counts and the faster clock speeds requiring more layers for noise suppression and signal integrity. Again the results are based on empirical observations rather than any mathematical derivations, but a general estimate for the number of layers (N_L) can be obtained using:

$$N_L = 5 \log[A_n f_{CLK}] \hspace{3cm} 7.20$$

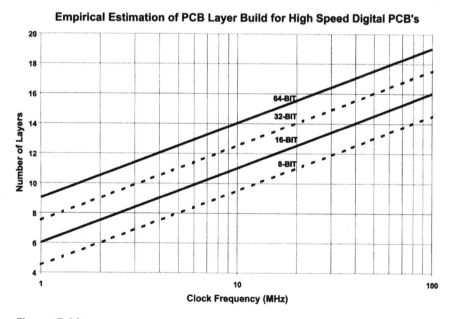

Figure 7.10

Number of layers in high-speed digital system PCB designs

This empirical equation should be used as a first approximation only and as an initial gauge of PCB cost. The number of layers suggested is usually slightly pessimistic and fewer layers should be possible.

The applicability of the equation can be examined by using it to estimate the number of layers in PCBs, which you already have worked on. Often the increased layer count suggested may not seem necessary, for example a 16-bit microprocessor system operating at 4 or 66 MHz will not really require any difference in tracking layers for functional performance, but the equation would suggest nine layers at 4 MHz and 12 layers at 66 MHz. These additional layers would be expected to be used to improve the EMC performance (i.e. additional ground planes) and to match the impedance of the faster signal tracking in the 66 MHz system (e.g. matched striplines).

7.2.2 Layer Build

The structures of layer build should be similar to the suggestions given for general PCB design, with possibly the inclusion of more ground and power planes and

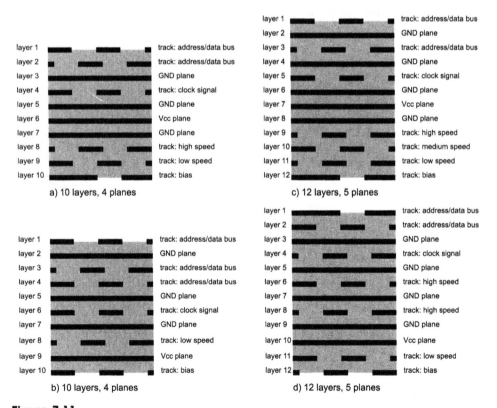

Figure 7.11

Example PCB stacks

tighter control over layer thicknesses. With more layers likely in higher-speed designs, it should be possible to dedicate specific layers to specific routeing and tracking functions (e.g. layer 1 for bias tracks, layer 3 for data buses, layer 5 for clock and timing tracks, etc.).

The use of foil build (i.e. prepreg layers between tracking) may be more frequently used in high-speed PCB designs, particularly between ground and power planes. The use of foil build on power planes produces a good high frequency distributed capacitor which improves decoupling with almost no cost implications. Foil build between tracks and ground is generally not recommended as this creates such a low impedance that the signal can be difficult to drive.

As there are going to be several layers of high-speed signals within the PCB there should be multiple ground planes within the structure. A target of one ground plane per two signals layers carrying frequencies above 50 MHz should be aimed for. This allows multiple offset stripline constructions to be arranged as well as surface and embedded stripline tracks. Biasing and lower speed signals can either be tracked on the surface or between a power and ground plane. At clock rates above 100 MHz more stringent transmission line tracking may be required and a ground plane per high-speed tracking layer may be required.

7.2.3 Line Spacing

Coupling between high-speed tracks running in parallel can become problematic if the signals are not deliberately paired (e.g. differential signal pairs). Although this effect is small compared with cross-over capacitance the coupling can be further reduced by observing a simple line spacing rule based on the track width (w).

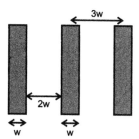

Figure 7.12

3w rule for track separation

The rule is sometimes referred to as the $3w$ rule as it suggests that track centres should be at least three times their width apart. This can also be applied to guard rings where the purpose of the ring is solely for radiated noise protection and deliberate coupling is not intended. The fringing field has fallen to 96% of its value immediately next to the track at this $3w$ distance.

Of course, where tracks are deliberately coupled, the converse would be true and tracks should be close. As a sensible limit a distance between centres of twice the track width provides adequate coupling between signals without risking short circuit problems. Again this can be applied to guard rings where some level of coupling to the guard rail and signal is desired.

7.3 Summary

At a seminar on portable systems design a US presenter made a comment about the enquiries they had on their technical helpline, '90% of all the reported problems were caused by noise, 80% of these were solved by layout changes'. I think a similar statement could probably be made about EMC problems within design, although I would doubt the statistics would be quite so high. What this statement does illustrate is the effect that PCB layout can have on a system's performance, both functional as well as for EMC compliance. Any design incorporating a PCB will not cost any more due to following the guidelines given here, but may cost a lot more if they are ignored.

Chapter 8

Software

There are two areas in which software can influence the electromagnetic compatibility (EMC) of a system or circuit. One is in the programming of microprocessors or controllers in digital systems and the other is in the use of design software to predict EMC performance. The fit of software into a book of this title may seem odd, but at the code level the software can be considered as another component in the system. At the design level there is little other than to review the available software techniques that can be covered within the scope of this book.

The usefulness of changing certain aspects of program code can quite easily be observed to improve the EMC performance of a circuit. The usefulness of some of the design software available is more difficult to assess as it is often difficult to know how something may have been done without the influence of the software.

8.1 Programming Issues for EMC

A program within a system, microcontroller or even a simple logic controller can have an influence on both the immunity and emissions of a system. Immunity is easiest to consider as many error checking and code validation routines already exist to ensure that corrupted transmissions are not accepted. Hence many programmers are already aware of the issues of software immunity, although they may not have considered this as an EMC issue.

On the emissions side it is more likely to be a case of acknowledging the best practices for addressing and operating the hardware from software codes. For example, setting a UART to tristate from input, then to outputs may reduce the transient power demand compared with a direct input to output change of state. This is generally known as defensive programming.

Many techniques, as with the watchdog circuit (see below and the section on ICs), do not actually improve emissions or immunity but simply provide a controlled method of recovery. Techniques involving simple known recovery states are usually less program intensive than performance improving techniques. The latter tend to have a higher programming overhead (i.e. require more code to execute) and more memory for storage.

8.1.1 Watchdog Programming

When incorporating a watchdog circuit (see IC section) into a design, some programming may be required for the timer reset of the watchdog. If possible the code should poll the watchdog pin once per program cycle. This is not too much of a problem with short microcontroller code or sequential routines, but with general operation processors and long programs this may be difficult. There may therefore need to be additional timed interrupts or hardware-generated interrupts to address the watchdog (e.g. dedicated watchdog interrupt handling code, Figure 8.1).

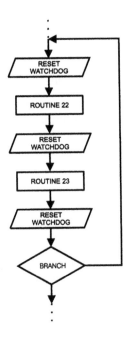

Figure 8.1

Sequential watchdog resets

The watchdog timer must be short enough to ensure non-catastrophic failure of the circuit and long enough not to interfere with functionality of the program (durations between 10 ms and 2 s are typical for watchdog timers). The trade-off in timing can be difficult to gauge correctly for some coding. Processors with a Harvard bus structure (independent data and instruction busses) are easier in this respect as each instruction is known to require a single clock cycle; therefore code timing is simple, count the instructions and divide by the clock frequency:

$$watchdog\ time\ out = \frac{number\ of\ instruction\ cycles}{watchdog\ clock\ frequency} \qquad 8.1$$

Another potential problem could occur if the watchdog code is not included within the sleep cycle for those processors with low power sleep modes. When the processor goes into sleep mode and the internal clock frequency is reduced, unless the watchdog interrupt or poll function is similarly adjusted, a reset could be implemented simply because the sleep mode was activated.

Watchdog programming has memory overhead as it requires additional program routines or interrupts. This may not be possible with fixed ROM microcontrollers and a hardware solution may be required. Likewise, in real time operating systems (RTOS), interrupts or software-based polling may not be possible due to their effect on program functionality. One simple hardware solution is to monitor lines that are known to change frequently, such as the lowest order address or data line, and use this for the watchdog polling line (Figure 8.2). The biggest problem here is ensuring that these lines will always change with variable code and within the watchdog timing regime. The advantage of hardware is that it requires no programming, therefore less memory. The disadvantage is the increase in component count to implement the function. It is most likely to cost more to implement the watchdog function in software than in hardware, but code is generally the best implementation, especially as a crashed program could continue to toggle the lines a hardware-based watchdog system is monitoring.

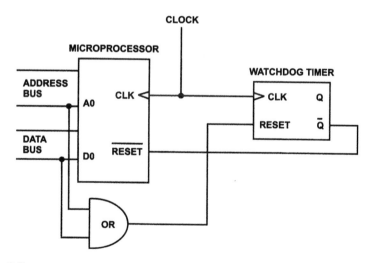

Figure 8.2

Hardware-generated watchdog interrupt

The most effective reset coding for a watchdog timer is to dedicate an output pin of the microprocessor to the watchdog (Figure 8.3). This pin should be set and cleared alternately as the program routines are executed (Figure 8.4). By using an AC coupled watchdog set and reset, if the code loops within a single routine that includes a watchdog set command, the watchdog is still effective and this single port condition status will operate the reset. This effectively doubles the robustness of the watchdog with only a small code overhead (alternating port set and port clear commands).

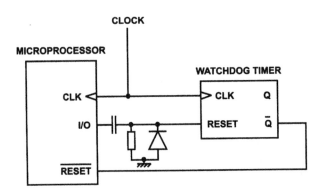

Figure 8.3
AC coupled watchdog circuit

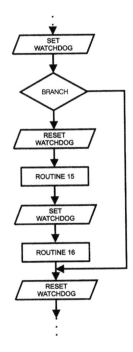

Figure 8.4
Watchdog set/reset routines

8.1.2 Refresh Port Connections

The data direction registers and the input/output port data registers are usually located near the edge of the processor's package if included in the microprocessor and may be connected directly to the external circuits. Consequently, these are highly likely to have noise on their lines. A simple way of minimising this noise disturbing the data settings and propagating into the microprocessor integrated circuit (IC) or system is to refresh these registers regularly.

The action of the microprocessor rewriting these data registers stabilises the interfaced circuits and minimises the risk that noise on these ports is corrupting other internal registers. This is a simple task to perform regularly and involves minimal programming overhead, just occasional write commands to the ports and data registers.

Some care does need exercising to ensure that writing the port status is appropriate to the program activity at each re-write command. It can not be assumed that the value in the port status is the correct setting so a read and then re-write could result in enforcing an erroneous setting.

8.1.3 Polling Interface Ports (Oversampling)

Input and output to interfaced functions usually occurs at a much slower rate than the microprocessor clock. This enables the microprocessor to poll these pins several times to ensure that either the level is set for an output or that the incoming signal is stable. This is analogous with the types of oversampling used in digital audio circuits (e.g. CD players).

The microprocessor can be programmed with an oversampling scheme, either to ignore the first and last and take the mean of a middle sample for instance, or simple average. The actual scheme for polling can be left to the programmer and will depend on the type of interface being addressed (e.g. audio, keyboard, serial data link) and the processing rate.

This type of programming has a major software and run time overhead as it requires several repeats of an operation plus usually some mathematical interpretation. The result is an improved immunity to noisy input signals and less susceptible output ports. This may best be implemented using a dedicated input/output microcontroller with this type of software programmed on-chip rather than as an auxiliary program for a main processor.

Dedicated input/output interface controllers are quite popular in larger systems hence adding local code to improve the immunity should be relatively easy. The same code could then be used for any system with that type of interface controller making the immunity improvement portable to other systems.

8.1.4 Token Passing

Token passing is a deliberate method of ensuring the program is progressing in a controlled manner and has not been jumped into due to a code corruption. This requires additional programming steps at each critical subroutine or program block within the final program.

Token passing requires a token (value) to be either updated in memory or in a register. As each routine is called it checks to ensure that the call is from the previous set of code by comparing the token with a programmed value or memory location. If a jump has occurred prematurely the token will not be set correctly, likewise if the call has come from a random corruption the token will be in the wrong state. On the discovery of a wrong token a reset routine is initiated.

The advantages of token passing over other methods are that the program continuously looks for errors rather than waits for an error to occur and the token itself can be used to reset to a predefined location in the program. It may not be necessary to completely reset the system depending on the location of the error and the token value, this obviously saves program time but also gives a genuine level of immunity rather than a simple reset on error recovery. The extra programming overhead is in adding the compare program to the start of each sub-routine and the token update to the end, plus any error handling software control if system reset is to be avoided.

```
Code              Comment
...
:sub21            start of sub routine
LDA     &token    load accumulator with token
LDB     &value    load correct token value
CMP               compare token and value
BNE     #reset    jump to reset routine if token incorrect
XXX               otherwise proceed with program
...
XXX
LDA     &token    load token
INC               increment token value for next routine
STA     &token    save token
:end_sub21        end of sub routine program
...
```

If attempting to use the token passing technique to recover to a known program position more programming effort will be required and probably several tokens to ensure register and flag conditions are suitable for re-entry into areas of code. Using the token value alone is insufficient to reset routines where register or memory writes have occurred as these will most likely be incorrectly set.

8.1.5 Unused Memory Addresses

In most applications it is impossible to exactly fill all memory locations, either ROM or RAM. These locations could be accessed due to a software or hardware addressing error caused by a corrupted code. These unused memory locations should therefore have a known operation placed in them, usually either a no operation (NOP) or an unconditional jump to a reset routine (JMP RESET).

```
Address       Code           Comment
0000          XXX            functional program area
...
00FA          XXX            end of program area
00FB          NOP            start of unused memory block
00FC          NOP
...
010F          JMP RESET      jump to reset routine
0110          XXX            functional program area
...
```

Where blocks of memory are unused, the jump to reset should be at the bottom of the memory with no operation commands above. These blocks may be scattered about the ROM/RAM locations and need to be carefully mapped if this type of feature is to be used.

Writing these areas will be required by the initialisation program if they are in RAM locations. There is of course the possibility that a corrupted write code could rewrite these NOP instructions in the RAM space, similarly with the program itself, but it is impossible to program for all eventualities and a reasonable trade between effectiveness and efficiency of programming has to be drawn.

This technique requires only a small amount of additional programming to the start-up (boot) routine and should not require any run-time program overhead.

8.1.6 Code Ghosting

This is one of the most memory intensive and possibly expensive techniques, usually reserved for fault tolerant software. The technique requires each code or data value to be stored in two parallel locations (either in identical or in complement form). The code or data are then compared with their ghost value prior to use.

The potential for electromagnetic interference (EMI) causing both codes to be corrupted is extremely low; however, the microprocessor still has no way of knowing which of the two codes is correct and which corrupt. The microprocessor would have to handle a code mismatch by reloading to check if the error was in the load operation or abort to a known routine if the error is in the stored memory values.

```
Code           Comment
...
:sub32         start of sub routine
LDA    &code   load accumulator with code value
LDB    &ghost  load ghost value
CMP            compare value and ghost
BNE    #error  jump to error routine if values do not match
XXX            otherwise proceed with program
...
```

This method is a hardware implementation of the oversampling technique and could be extended if necessary, although at some cost if there is a lot of code (i.e. three or four copies could be used for comparison). If multiple ghost values are used a mathematical method of selecting the correct value could be applied (e.g. numerical digital average of each bit). The method can also be selectively applied to code or data values from a known problematic location or EMC critical area of the system or program.

8.1.7 Other Techniques

There are other techniques that are more specific to the hardware being used and the best methods of operating interfaces and associated circuits. For example the interface with a tristate setting mentioned in the first part of this section. If a setting of the port from logical outputs to inputs causes a large switch in internal states of an IC, resulting in a large current demand, but an intermediate state is available that has only a fraction of the transient current demand when switched (Figure 8.5). Using the intermediate state between transitions requires only one additional code instruction and will reduce transient supply demand and hence conducted noise levels.

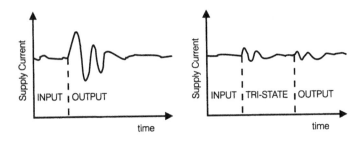

Figure 8.5

Effect of port status changes on supply current

The intermediate state transition technique may be applicable to other circuits, such as clearing selective flags prior to rewriting may save the number of actual data or address lines driven and therefore again reduce transient demand. Actual transient savings in many circuits will be negligible and the idea is best reserved for those functions which are known to require larger current supply (e.g. line drivers, interface circuits and bi-directional ports).

Another technique that can help with interfaces is to use a coding scheme, such as Manchester coding, which has only a few frequencies in the transmission. In Manchester coding only two frequencies are used, a '1' is indicated by a single transition and a '0' by a double transition within the clock cycle (Figure 8.6). The signal is therefore always changing state, hence a latched condition or end of transmission are also easy to detect. The receiving circuit can potentially determine if the sender is experiencing EMC problems. With standard non-return to zero (NRZ) coding the frequencies present in the transmission can be from the clock frequency to whatever the maximum bit code length period is (i.e. 1/8 the clock frequency for an eight-bit code, Figure 8.7).

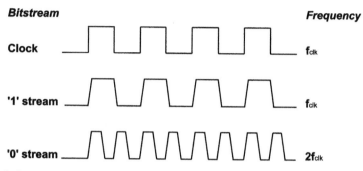

Figure 8.6

Manchester code

Manchester coding makes filtering easier, if required, and allows for a simpler method of triggering on levels rather than edges that will also improve immunity. The clock of the transmission circuit is derivable from the signal hence, using a fixed offset delay, a level triggered receive circuit is easy to implement in either software (using a delay routine) or in hardware (using a delay line).

Converting standard edge triggered circuits to level triggered is possible at specific ports by using a software code delay to allow the value to settle prior to reading. The

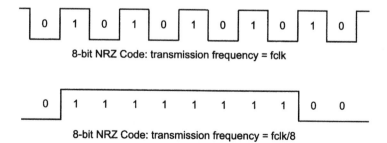

Figure 8.7

NRZ coding

length of delay will depend on the function, but should allow the ringing or overshoot to settle and consequently not affect the value read at the port (Figure 8.8), a typical settling time should be 10% of the maximum clock period. This can also be implemented in hardware using a delayed latch trigger with the data ready signal coming from the delayed latch set condition (Figure 8.9).

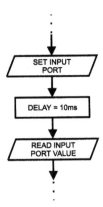

Figure 8.8

Software delayed port read

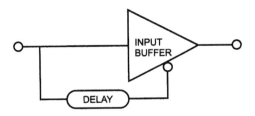

Figure 8.9

Hardware delayed signal

8.2 Design Software

Software tools for design for EMC tends to fall into two distinct categories of program, one is the simulator the other the advisor. The simulator attempts to predict precise values of field or conducted noise at any chosen point in a circuit system or structure. The advisor usually consists of a knowledge base and is applied to an otherwise complete design to check its compliance with the rules of the knowledge base.

The effectiveness of the software is difficult to gauge accurately without comparison with measured results and unfortunately little comparative testing with finished circuits is recorded. Most of the packages currently available concentrate on the physical side of circuit design, being concerned with the dimensional construction of PCBs and component packages. To date simulation of circuit behaviour and component performance is still performed where possible on existing standard circuit simulators (e.g. SPICE).

8.2.1 PCB Design Software

PCB design software is available from a bewildering number of suppliers with a long list of various features of each package to differentiate it from its rivals. This is usually one area of design that is almost exclusively done by computer, even the simplest of circuits, designed on the back of an envelope, bread boarded rather than simulated, is still usually laid out for production in a PCB design package.

The PCB design software does allow the designer to control more of the EMC performance than component selection, as the designer can choose the dimensions of interconnect, the number of layers present and their use. One feature that is now prevalent on PCB design packages that is a potential problem for EMC is the auto-router. The PCB design software does not necessarily know which are the fastest signals and therefore which require priority in the layout, consequently initial clock and high speed tracking may have to be done manually with the auto-router used only for the bias components, low speed and DC tracks.

The major thrust for advisor type software for EMC is in the area of PCB design and these can often operate on prelaid designs and give an indication of likely problem

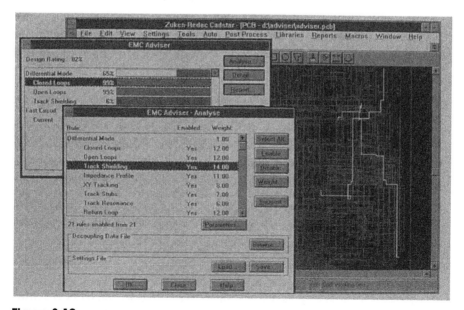

Figure 8.10

EMC Advisor Analysis Software System from Zuken-Redac

areas. Often the advisor may need some considerable input from the user to establish some rules of layout specific to the user's applications. The amount of input can be tedious and many advisor programs are little more than an additional design rule checker (DRC) added to the existing DRC program. In fact if your PCB software has an edit function for its DRC it is possible to create your own EMC advisor.

Simulation software at the PCB level takes the physical dimensions of the tracks and interconnect (again coupled with data on the PCB material, see Appendix B) to construct a transmission line model or aerial model for interconnect. This usually then requires the component model data to produce some idea of likely delays, propagation characteristics, termination values, etc. to predict if the layout will cause functional problems (usually determined by signal integrity) or EMC problems. Usually, the software is a conventional circuit simulator (electrical for conducted noise and electromagnetic (EM) for radiation) and the PCB data are extracted from a PCB layout package to produce the relevant interconnect models only. This, therefore, can require integration of software packages: a PCB layout package, a PCB interconnect model extraction tool, a circuit simulator and EM radiation modelling package.

As with any simulation the result is only as good as the models allow. In other words the data on the PCB characteristics and dimensions have to be accurate for the supplier you use. Another problem with this type of simulation is the time required for extraction and analysis. Consider a small PCB with 10 ICs in 14-pin packages, there are at least 140 interconnections as well as the components to simulate. Consequently, these simulation models can usually only be run on the few critical interconnects to check integrity and functionality within a reasonable simulation time, especially if the radiation is being examined.

The above simulator arguments generally apply to predicting emissions for which some models do exist. The susceptibility is, however, much more difficult, little component data are available and the EM software is highly complex for such tasks. In general, the assumption that low emission designs offer low susceptibility is usually used.

8.2.2 Component Simulation Software

At the component level electrical circuit simulators have been available for a long time and several *de facto* standard packages exist. For electrical simulation of analogue circuits SPICE has gained a wide acceptance and many examples of its accuracy exist as do many models of passive, discrete and integrated components. Digital circuit simulators are also common, but do not lend themselves to encompass the possibility of noise analysis for EMC as an analogue simulator does. Consequently, extensions to digital circuits to allow them to interface to analogue or mixed mode simulators is now a relatively common theme, also digital models in SPICE are available.

The main problems are that the EMC performance is often outside the normal operating range of the components, and hence their model. Some models do exist but component packaging models and interconnect (see section on PCB software) also have a significant effect on the EMC performance and models for these are not quite as common.

An area in which component simulation lacks behind PCB simulation is in emissions modelling. The component package is not usually within the control of the circuit designer, they require the device's function not its package. Consequently, unless the component supplier can provide package data (dimensions of tracking and bonding as well as electrical parasitics), modelling emissions from components is not feasible for a component user. Even the supplier will experience difficulty and the use of any model derived would be very limited as few people are attempting this level of modelling or have the time or facilities to do so. The EM software that can perform this level of modelling is expensive and requires some expertise to operate and a significant computing performance from its host platform.

An emerging modelling standard that will assist in the conducted emissions simulation is IBIS (Input/output Buffer Information Specification), which includes package electrical parasitic data and rise and fall time information. The device models are relatively basic in their functional performance within the IC, but give reasonably detailed information of the behaviour of the signals at the pins of the device. The models offer one of the best solutions to mixed signal simulations as well as offering possible EMC simulation data and without compromising the IC manufacturers' circuit details. As with any model there are limitations, and therefore due to the simplicity of the model it is unlikely to predict accurately reaction to incident phenomena, such as high frequency conducted input noise signals or accurately model transients impact behaviour (although some transient absorption models are included in some devices) and EM emissions are not modelled.

8.2.3 Design Software Overview

There is still a gap between component and PCB design software, although both are commonly used by circuit designers, they are considered disparate operations in the design cycle. If software is going to be used to predict circuit level EMC performance it is going to have to link the component, circuit and PCB information.

Some software vendors are offering the possibility of taking PCB files and adding models to a circuit simulator to model interconnect. There does need to be some feedback if the PCB layout is then going to determine which tracks carry the highest speed signals, therefore which to prioritise for layout. The simulator link approach requires collaboration between simulation and PCB CAD vendors, this is already occurring as the ECAD vendor base reduces by merger and acquisition. The main drawback with simulator linked PCB layout software is the cost and complexity of linking these packages together; each has a primary function which is not necessarily compatible with the others'. This applies to both the analogue and digital circuit simulators currently available as well as PCB layout packages.

Another consideration for the ECAD supplier is are designers willing to pay for EMC add-ons to their software and are they willing to trust the results? No original equipment manufacturer (OEM) would go to market with a product simulation and no test data. As testing will still be required the value of a simulation tool will be based on a trade-off between convenience, accuracy and of course cost.

The value of simulation should be in reducing the design time. This requires the correct set of design rules and models within the simulator and the correct application to both circuits and systems. As in the early days of analogue and digital simulations, the loop closing the simulation versus measured performance still needs completing for EM software to enable designers to believe in the results the software provides. This is being performed and the software is continually being refined. Eventually the simulation will provide adequate information on EMC performance and tests will be performed solely as verification rather than an iterative process in the design cycle.

8.2.4 Available Commercial Packages

The following tables (Tables 8.1–8.4) list some of the available software packages aimed at EMC simulation or modelling. There are no doubt others that offer various features and new packages are appearing each year. There is no best solution as it will depend upon many factors, not least of all the price of the software.

Table 8.1 *Advisor-based software*

Product name	Supplier	Application
Design Advisor	Zuken-Redac	PCB DRC
EMC Expert Consultant	Seaward Ltd	General
EMC Toolkit	Continental Compliance	General
UniSolve	UniCAD	PCB DRC

Table 8.2 *Conducted emissions modelling software*

Product name	Supplier	Method	Application
em	Sonnet Software	MOM	circuits semiconductors
EMA3D	Electro Magnetic Applications	FDTD	3D structures
EMC Workbench	Incases	TLM/MOM	PCB design
EMIT	Altium	MOM/FDTD	conductors
Greenfield 2D	Quantic Laboratories	TLM	signal integrity PCB design
L-Edit/EM	Tanner Research	BEM	PCB layout
Maxwell Strata	Ansoft	MOM	signal integrity, cross-talk
Micro Stripes	Kimberley Communication Consultants Ltd	TLM	3S structures antennas
Motive	Quad Design	TLM	PCB layout
Superstar	Eagleware	TLM	PCB design

Table 8.3 *Radiated emissions modelling software*

Product name	Supplier	Method	Application
ContecRADIA	Contec Microelectronics	Antenna Mathematics	PCB traces, wires and structures
Greenfield 3D	Quantic Laboratories	BEM	3D structures and PCB traces
MAFIA	Computer Systems Technologies	FDTD	3D structures
Maxwell SI Eminance	Ansoft	FEM	3D structures
Momentum	HP-EEsof	MOM	planar structures
MSC/EMAS	MacNeal-Schwendler Corporation	FEM	3D structures, antennas, EMI and cross-talk
QUIET	Quad Design	MOM	Any 2D and 3D structures

Table 8.4 *General electromagnetic modelling software*

Product name	Supplier	Method	Application
HFSS	HP-EEsof	FEM	3D structures
Maxwell Eminence	Ansoft	FEM	3D structures
OERSTED	Integrated Engineering Software	BEM	time harmonic electromagnetic fields
TOSCA	Vector Fields	FEM	3D structures

There are many ways to handle EM simulation, the most complex is the finite element method (FEM) which is commonly used in mechanical stress design software. The EM versions solve Maxwell's equations in space along a grid or mesh either defined by the user or by the software. FEM is arguably the most accurate method but also the most computationally intensive and the most expensive. FEM can be difficult to use but is very versatile and can be applied to almost any structure. The boundary element method (BEM) is similar but deals with boundary and space conditions rather than within finite elements of a grid, it is therefore a little faster and of similar accuracy to FEM. The transmission line matrix (TLM) is one of the faster methods that attempts to create two- or three-dimensional transmission lines in space and calculate fields at various points based on transmission line equations. The TLM method is fast but not as accurate as BEM or FEM for radiated emissions.

Other methods exist for solving the radiated fields or conducted noise within circuits and structures. They all offer some feature that makes them different from the next, but usually each has a trade-off in speed, price, accuracy and versatility. Although FEM seems the 'best' solution, there are certain applications where the other techniques can yield results of similar accuracy with faster computation times and at a lower cost in both software and in time spent on the model construction. For arbitrary geometries and non-linear materials FEM may be the only suitable method. As with any simulator, their accuracy ultimately lies with the accuracy of the models contained within the software or constructed by the user.

A list of the software vendor addresses is given in Appendix C.

Appendices

Appendix A

EMC Standards

The standards field for electromagnetic compatibility (EMC) is one of the first areas to be covered in the EU by what are known as 'new approach directives'. This means that many of the product specific directives are in fact just guidelines to application, the appropriate testing methods and limits may be contained in other European Norme (EN) standards. It would not be uncommon to find that it requires several EN standards to encompass fully the testing required on a single product item. There is no straightforward way to navigate through these standards, it will undoubtedly take time and care, but with experience the designer will learn what standards apply to their designs and be able to list them in the test schedule.

A fuller explanation of the standards is contained in Chris Marshman's book, *Guide to the EMC Directive*. The tables listed here are for first time guidance only and are not a definitive list. The EC is also revising and updating the standards relatively regularly on EMC, hence some may have been superseded or replaced. A good point of contact for the most up-to-date information is one of the standards issuing bodies (e.g. BSi, VDE, SEMKO, FCC).

Information technology equipment (ITE) *(including Telecommunication Terminal Equipment)*

Application	Emissions	Immunity
Conducted RF immunity		prEN 55024–6 IEC 801–6
Electrostatic discharge		prEN 55024–2 IEC 801–2
Fast transients/bursts		prEN 55024–4 IEC 801–4
Generic	EN 60555–2	EN 50082–1 prEN 55024–1
ISDN	prEN 55012–1	prEN 55102–2
50 Hz magnetic fields		prEN 55024–7
Radiated RF immunity		prEN 55024–3 IEC 801–3
Radio interference characteristics	EN 55022	
Supply disturbances/voltage fluctuations	EN 60555–3	
Surges		prEN55024–5 IEC 801–5

Residential, commercial and light industrial equipment

Application	Emissions	Immunity
Alarm systems (without mains connection)	EN 50081–1	
Audio, visual and audio-visual equipment for domestic entertainment	EN 60555–2/3 EN 55013	EN 55020
Building automation	EN 60555–2/3	
Broadcast satellite receivers		
Conducted and radiated emissions	EN 55022	
Domestic appliances and small household appliances (including toys)	EN 60555–2/3 EN 55014	
Electronic switches	EN 60555–2/3 EN 55014	IEC 669
Generic	EN 50081–1	EN 50082–1
Lifts and low voltage motor drives (domestic)	EN 60555–2/3 EN 55014	
Lighting, fluorescent lamps and luminaires	EN 60555–2/3 EN 55015	
Low voltage circuit breakers and similar	IEC 947	
Mains signalling at low voltage	EN 50065–1	prEN 50065–2
Residual current devices		
Small power electronics (power supplies), stabilised power supplies with DC output	EN 6055–2/3 IEC 478–3/5	

Special

Application	Emissions	Immunity
Electrical and electronic test and measuring instruments	EN 60555–2/3	
Industrial, scientific and medical equipment (ISM)	EN 60555–2/3 EN 55011	prEN 50097

Traffic and transportation equipment

Application	Emissions	Immunity
Electrical installation of ships		
Electric traction equipment	IEC 571–1/3	
Electronic equipment used on rail vehicles		
Motorway communication and traffic control equipment	EN 60555–2/3	
Navigational instrumentation	IEC 945	

Utilities

Application	Emissions	Immunity
High voltage fuses	EN 50081–1	EN 50082–1
High voltage switch and control gear		
Measuring, metering and load control apparatus (electronic)	IEC 1036 IEC 1037 IEC 1038	
Protection equipment		IEC 225–22–2/4
Public low voltage power supplies signalling	prEN 61002–2	
Telecontrol, teleprotection and associated telecommunications for utilities		

Appendix B

Printed Circuit Board (PCB) Design Parameters

Many PCB manufacturers still use imperial or mixed imperial and metric units. In the USA imperial is still the preferred choice of units for almost all applications. Both metric and imperial units are quoted here, although some approximations are made between the two. The units are determined from values given in one unit of measure and converted using 25.4 mm = 1 inch conversion factor.

Laminate Material Properties

Base material	Reference name	Relative dielectric constant ε_r	Maximum temperature T_{max} (°C)	Thermal conductivity k (W/m per K)
Difunctional epoxy	FR4	4.2–4.9	120–130	0.18
Tetrafunctional epoxy	FR4	4.4–4.9	>135	0.18
Multifunctional epoxy	FR5	4.2–4.9	140–170	
Alumina ceramic	Al_2O_3	10		20
Polyimide		3.5	260	
Teflon		2.2		

Values of relative dielectric for laminates can vary with the manufacturer. With FR4 it is dependent on the glass fibre to epoxy resin mix. A typical value of 4.7 is used throughout this book.

Copper track properties

Copper weight (oz)	Track thickness		Track resistance (mΩ/□)
	mm	inches	
0.5	0.018	0.0008	0.94
1.0	0.035	0.0015	0.49
2.0	0.070	0.0030	0.24

The resistance figure is in milliohms per square (mΩ/□). This is the resistance of any single square section of track. For a track which is 1 mm wide and 10 mm long, this would contain 10 square sections, hence for a 0.5 oz copper layer would have a resistance of 9.4 mΩ. These resistance values are based on an electrical conductivity value of 17×10^{-9} Ωm for copper.

The most common copper weight in use is 1 oz and is used for the examples in this book.

Standard FR4 laminate and prepreg thicknesses

Laminate		Prepreg		
mm	inches	Code	mm	inches
0.100	0.004	116	0.050	0.002
0.115	0.0045	1080	0.065	0.0025
0.128	0.005	2113	0.075	0.003
0.150	0.006	2125	0.105	0.0042
0.200	0.008	7628	0.180	0.007
0.250	0.010	7648	0.190	0.0075

The table (standard FR4 laminate and prepreg thicknesses) shows some of the smaller commercial standards sizes. Thicker laminates can be constructed from either multiples of the above or combinations of laminate and prepreg. There is also a range of 'off-the-shelf' sizes in 0.05 mm (0.002") steps for both laminates and prepreg available from some manufacturers.

Minimum track and gap dimensions and probable yield

Track width		Gap		Yield (%)
mm	inches	mm	inches	
0.20	0.008	0.20	0.008	96
0.15	0.006	0.20	0.008	94
0.15	0.006	0.15	0.006	90
0.12	0.005	0.15	0.006	99
0.12	0.005	0.12	0.005	83
0.10	0.004	0.12	0.005	75
0.10	0.004	0.10	0.004	50

Fine tracks and gaps generally produce a yield loss at the PCB supplier. This yield loss is passed back to the customer in the form of increased board costs for fine geometry PCB designs. The table (minimum track and gap dimensions and probable yield) gives approximate yield losses for fine line track and gap sizes and may be useful to decide a minimum dimension without relying on the supplier's minimum track capability, which will include a premium.

Minimum drilled hole Size and PCB Thicknesses

PCB thickness		Drilled hole size		Stack height
mm	inches	mm	inches	
<1.6	0.065	<0.2	0.008	1
<1.6	0.065	0.2–0.3	0.008	2
<1.6	0.065	0.3–0.5	0.012–0.020	3
<1.6	0.065	>0.5	>0.020	4
1.6–2.4	0.065–0.095	<0.5	<0.020	1
1.6–2.4	0.065–0.095	>0.5	>0.020	2
2.4–3.2	0.095–0.125	<0.5	<0.020	1
2.4–3.2	0.095–0.125	>0.5	>0.020	2
>3.2	>0.125	any	any	1

The data quoted here is taken from a variety of sources, including manufacturers' data sheets, and is kept deliberately brief. The best source of these data for any given application is the manufacturer of the PCB. More detailed standard PCB design data of this sort can be obtained in the document ANSI/IPC-D-275 Design Standard for Rigid Printed Boards and Rigid Printed Assemblies.

Appendix C

Software Vendor Addresses

Ansoft Corporation, Four Square Station, Suite 660, Pittsburgh, PA 15219–1119, USA (Ansoft Europe, Regal House, Ninth Floor, 70 London Road, Twickenham, TW1 3QS)

Computer Simulation Technology GmbH, Lauteschlagerstrasse 38, D64289 Darmstadt, Germany

Contec Microelectronics, 4000 Moorpark Avenue, Suite 209, San Jose, CA 95117, USA

Continental Compliance Ltd, 4 Newlands House, Newlands Science Park, Hull, HU6 7TQ, UK

Eagleware Corporation, 1750 Mountain Glen, Stone Mountain, GA 30087, USA

Electro Magnetic Applications Inc, 7655 West Mississippi Avenue, Suite 300, Lakewood, CO 80226, USA

HP-EEsof, Hewlett-Packard Test & Measurement, 5301 Stevens Creek Boulevard, Building 51L-SC, Santa Clara, CA95052, USA

Incases Engineering GmbH, Vattmanstrasse 3, 33100 Paderborn, Germany

Integrated Engineering Software, 46–1313 Border Place, Winnipeg, Manitoba, R3H 0X4, Canada

Kimberley Communication Consultants Ltd, 104 GPT Business Park, Beeston, Nottingham, NG9 2ND, UK

MacNeal-Schwendler Corporation, 815 Colorado Boulevard, Los Angeles, CA 90041–1777, USA

Quad Design, 1385 Del Norte Road, Camarillo, CA 93010, USA

Quantic Laboratories, 12th Floor, 191 Lombard Avenue, Winnipeg, Manitoba, R3B 0X1, Canada

Seaward Electronics Ltd, Bracken Hill, South West Industrial Estate, Peterlee, County Durham, SR8 2SW, UK

Sonnet Software Inc, 1020 Seventh North Street, Suite 210, Liverpool, NY 13088, USA

Tanner Research, 180 North Vinedo Avenue, Pasadena, CA 91107, USA

UniCAD Inc, 174 Littleton Road, Suite 103, Westford, MA 01886, USA

Vector Fields Ltd, 24 Bankside, Kidlington, Oxford, OX5 1JE, UK

Zuken-Redac, Green Lane, Tewkesbury, Gloucestershire, GL20 8HE, UK

GLOSSARY OF TERMS

Abbreviations and Acronyms

ADC	analogue to digital converter
ASIC	application specific integrated circuit
BCI	bulk current injection
BEM	boundary element method
BiCMOS	bipolar and CMOS
BGA	ball grid array
BJT	bipolar junction transistor
BNC	Bayonet Neill-Concelman
CAD	computer-aided design
CD	compact disk
CMOS	complementary metal oxide semiconductor
CMRR	common mode rejection ratio
DAC	digital-to-analogue converter
DIL	dual-in-line
DRAM	dynamic random access memory
DRC	design rule checker
DSO	digital storage oscilloscope
EC	European Community
ECAD	electronic computer-aided design
ECL	emitter coupled logic
EDA	electronic design automation
EFT	electrical fast transient
EM	electromagnetic
EMC	electromagnetic compatibility
EMF	electromotive force
EMI	electromagnetic interference
EN	European Norme
ESD	electrostatic discharge
ESL	equivalent series inductance
ESR	equivalent series resistance
FATID	frequency, amplitude, time, impedance and dimensions
FDTD	finite difference time domain
FEM	finite element method
FET	field effect transistor
FEXT	far end cross-talk
FPGA	field programmable gate array

GDT	gas discharge tube
GTEM	GHz transverse electromagnetic
GTL	gunning transistor logic
HD	high density
HM	hard metric
IBIS	input/output buffer information specification
IDC	internal data channel
IC	integrated circuit
IGBT	insulated gate bipolar transistor
IRQ	interrupt request
ISM	industrial, scientific and medical equipment
ITE	information technology equipment
JFET	junction field effect transistor
LDO	low drop out (regulator)
LED	light-emitting diode
LVD	low voltage directive
LVDS	low voltage differential signalling
MLCC	multilayer ceramic chip capacitor
MLV	multilayer varistor
MOM	method of moments
MOSFET	metal oxide semiconductor field effect transistor
MOV	metal oxide varistor
NEXT	near end cross-talk
NRZ	non-return to zero
NRZI	non-return to zero inverted
OEM	original equipment manufacturer
OTP	one time programmable
PC	personal computer
PCB	printed circuit board (also printed wiring board)
PCMCIA	Personal Computer Memory Card Industry Association
PECL	pseudo emitter coupled logic
PGA	pin grid array
PLCC	plastic leadless chip carrier
PLA	programmable logic array
PLL	phase locked loop
PSDC	power supply damping circuit
PSU	power supply unit
PTH	plated through hole
PWB	printed wiring board (PCB used here)
PWM	pulse width modulation
QFP	quad flat pack
RAM	random access memory
RF	radio-frequency
RFI	radio-frequency interference
ROM	read only memory
RTOS	real-time operating systems

SCI	scalable coherent interface
SCR	silicon controlled rectifier
SI	signal integrity
SM	surface mount
SMD	surface mount device
SMPS	switched mode power supply
SMT	surface mount technology
SPICE	simulation program with integrated circuit emphasis
SRAM	static random access memory
SRF	self-resonant frequency
SSC	spread spectrum clock
SSR	solid state relay
STP	shielded twisted pair
TAB	tape automated bonding
TBU	transient blocking unit
TCF	technical construction file
TEM	transverse electromagnetic
TLM	transmission line matrix
TTL	transistor transistor logic
TVS	transient voltage suppressor
UART	universal asynchronous receive and transmit
UHF	ultra high frequency
UTP	unshielded twisted pair
VHF	very high frequency
VCO	voltage-controlled oscillator
VDR	voltage-dependent resistors
VME	virtual memory extended
ZIF	zero insertion force
ZCS	zero current switching
ZVS	zero voltage switching

International Standards Bodies and Institutions

ANSI	American National Standards Institute
BAPT	Bundesamt für Post und Telekommunikation (Germany)
BSI	British Standards Institute
CCITT	International Telephone and Telegraph Consultative Committee
CEI	Italian Electrotechnical Committee
CENELEC	Comite European de Normalisation Electrotechnique
CISPR	Comite International Special des Perturbations Radioelectriques
CSA	Canadian Standards Authority
DEMKO	Dansk Standards (Denmark)
DIN	Deutsches Institut für Normung (Germany)
DKE	Deutsches Elektrotechnische Kommission in DIN und VDE (Germany)
DTI	Department of Trade and Industry (UK)
EIA	Electronic Industries Association
ETCI	Electrotechnical Council of Ireland
ETSI	European Telecommunications Standards Institute
FCC	Federal Communications Commission (US)
HDTP	Hoofdirectie Telecommunicatie en Post (Netherlands)
IEC	International Electrotechnical Commission
ISO	International Organisation for Standardization
ITU	International Telecommunication Union
JISC	Japanese Industrial Standards Committee
NAMAS	National Measurement Accreditation Service (UK)
NEMKO	Norsk Elektroteknisk Komite (Norway)
NSF	Norges Standardiseringsforbund (Norway)
OVE	Osterreichisher Verband für Electrotechnik (Austria)
SAA	Standards Australia
SCC	Standards Council of Canada
SFS	Suomen Standardoimislitto (Finland)
SEMKO	Svenska Elektriska Kommissionen (Sweden)
SESKO	Finnish Electrotechnical Standards Organisation
SEV	Schweiz Elektrotechnisher Verein (Switzerland)
SIRIM	Standards and Industrial Research of Malaysia
UL	Underwriters Laboratory (USA)
UNI	Ente azionale Italiano di Unificazione (Italy)
VCCI	Voluntary Council for Interference by Data Processing Equipment and Electronic Office Machines (Japan)
VDE	Verband Deutsche Electrotechniker (Germany)

REFERENCES

This book is purposely aimed at the component and printed circuit board level, there are many other books on the market for the systems level and for those involved in electromagnetic compatibility (EMC) testing. There are also a significant number of seminars, colloquia and international conferences which cover all aspects of EMC.

The enclosed list gives most of the reference texts which were consulted in the compilation of this book. Although all have merit in their own right there are two I would recommend outright for their subject coverage. First, Tim Williams book *EMC for Product Designers* is the best overall coverage of the EMC subject at the complete product level. For general low noise advice the Henry Ott book *Noise Reduction Techniques in Electronic Systems* is still a leading text in the field.

Books

EMC for Product Designers, Tim Williams, Butterworth-Heinemann, 1992.

Noise Reduction Techniques in Electronic Systems, 2nd edn, Henry W. Ott, John Wiley & Sons, 1988, ISBN 0-471-85068-3.

Printed Circuit Board Design Techniques for EMC Compliance, Mark I. Montrose, IEEE Press, 1996, ISBN 0-7803-1131-0.

Guide to the EMC Directive 89/336/EMC, 2nd edn, Chris Marsham, E.P.A. Press, 1995, ISBN 0 9517362 7 2.

Electromagnetic Compatibility, Jasper Goedbloed, Prentice Hall, 1990, ISBN 0-13-249239-8.

Electromagnetic Compatibility in Power Electronics, Laszlo Tihanyi, Butterworth-Heinemann, 1995, ISBN 0-7803-0416-0.

Transmission and Propagation of Electromagnetic Waves, K.F. Sander & G.A.L. Reed, 1978, Cambridge University Press, ISBN 0-521-2192-X.

Controlling Radiated Emissions by Design, Michel Mardiguian, Van Nostrand Reinhold, 1992, ISBN 0-442-00949-6.

Controlling Conducted Emissions by Design, John C. Fluke, Van Nostrand Reinhold, 1991, ISBN 0-442-23904-1.

Capacitance, Inductance and Crosstalk Analysis, Charles S. Walker, Artech House, 1990, ISBN 0-89006-392-3.

Physical Models for Semiconductor Devices, John E. Carroll, Arnold, 1974, ISBN 0-7131-3307-4.

EDN's Designers Guide to Electromagnetic Compatibility, Daryl Gerke and Bill Kimmel, Cahners.

Analysis and Design of Integrated Electronic Circuits, Paul M. Chirlian, Harper & Row, 1982, ISBN 0-06-318214-9

Electromagnetics, John D. Kraus and Keith R Carver, McGraw-Hill, 1973, ISBN 0-07-035396-4.

Transmission and Propagation of Electromagnetic Waves, K.F. Sander and G.A.L. Reed, Cambridge University Press, 1978, ISBN 0-521-29312-X.

The EMC Handbook, Nutwood UK, March 1996.

Trade Journal Articles

A Practical Strategy for EMC Design. *IEE Electronics and Communications Journal*, Peter Butler and Stuart Withnall, IEE, August 1993.

Microcontrollers, Industrial Control Circuits and EMC, Peter Gare, *Electronic Product Design*, March 1994.

EMI Performance Comparisons of Distributed and Discrete Capacitors, Mason Hu and James Howard, *Compliance Engineering* European Edition, April/March 1995.

Avoid EMI Woes in Power-Bus Layouts, Dave Miskell, *Electronic Design*, November 6, 1995.

Don't Let Rules of Thumb Set Decoupling Capacitor Values, Vincent Grebb and Charles Grasso, *EDN*, September 1, 1995.

Techniques for Controlling Radiated Emissions, Stephen Lum, *Compliance Engineering* European Edition, July/August 1995.

Design at the Circuit Board Level is Critical to EMC Success, Ken Webb, *Assessment Services*, Approval, July/August 1995.

Low Skew Clock Drivers, Gary Tharalson, *Electronic Products*, May 1992.

Differential Signal Transmission Through Backplanes and Connectors, Mary Ann Fusi and Fabrizio Zanella, *Electronic Packaging and Production*, March 1996.

Use Simulation to Spot and Fix EMI Problems, *Electronic Design*, July 22, 1996.

High Performance Connectors – The Often Underestimated Weak Link, Roland Modinger and Michael Munroe, *Electronic Design*, February 3, 1997.

Conferences, Seminars and Colloquia

EMC in High Integrity Digital Systems, IEE Colloquia, 17 May 1991, 1991/104.

EMC Workbook, DTI, Findlay Publications, 1993.

Does Electromagnetic Modelling Have a Place in EMC Design?, IEE Colloquia, 4 February 1993, 1993/028.

Case Studies in EMC, IEE Colloquium, 22 April 1993, 1993/091.

Euro-EMC Conference, Reed Exhibition Companies (UK), October 1993.

Electromagnetic Hazards to Active Electronic Components, IEE Colloquium, 14 January 1994, 1994/008.

Predicting and Assuring EMC in the Power Electronics Arena, IEE Colloquium, 15 February 1994, 1994/063

Euro-EMC Conference, Reed Exhibition Companies (UK), October 1994.

Component Selection for EMC, Martin O'Hara, Thames Valley EMC Club, January 1995.

The Five Golden Techniques for PCB Design for Lowest Cost EMC Compliance, Keith Armstrong, Thames Valley EMC Club, 5 July 1995.

Euro-EMC Conference, Reed Exhibition Companies (UK), October 1995.

Controlled Impedance Circuit Boards and High Speed Logic Design, IPC-2141, April 1996.

Euro-EMC Conference, Reed Exhibition Companies (UK), October 1996.

Corporate Literature

EMC Electromagnetic Compatibility, Schaffner, 1985.

Transmission Line Effects in PCB Applications, Motorola Semiconductor Application Note AN1051/D, 1990.

EMC Design Guidelines, Newport Components Ltd, 1993.

A/D & D/A Converter Applications Seminar, Crystal Semiconductor Corporation, 1993.

The Bypass Capacitor in High Speed Environments, Texas Instruments, 1993.

EMC Guide, Lutze Systematic Technology, 1994.

A Multilayer Approach to Transient Voltage Suppressors, AVX Corporation.

Transient Voltage Suppression Device, Harris Semiconductors, 1995.

Printed Circuit Board Layout for Improved Electromagnetic Compatibility, Texas Instruments, 1995.

Designing for Electromagnetic Compatibility with Single Chip Microcontrollers, Motorola Semiconductor Application Note AN1263/D, 1995.

RF Connector Application Guide, Vitelec Electronics Limited, 1997.

Private Communications

Steve Jones, Manchester Circuits Ltd, Victoria Lane, Manchester, M45 6BU, UK.

Brian Smith, D2D Ltd, West Avenue, Stoke-on-Trent, ST17 1TI, UK.

Advanced Interconnection Technology, 3289 Montreal Industrial Way, Tucker, GA 30084, USA.

Neil Chadderton, Zetex PLC, Fields New Road, Oldham, OL9 8NP, UK.

Geoff White, EMC Net, http://www.emcnet.com.

Eric Bogatin, Ansoft Corporation, 4320 Stevens Creek Boulevard, Suite 163, San Jose, CA 95129, USA.

INDEX

Printed and bound by CPI Group (UK) Ltd, Croydon, CR0 4YY

07/10/2024

01041773-0001